REED'S
MOTOR ENGINEERING
KNOWLEDGE
FOR
MARINE ENGINEERS

REED'S
MOTOR ENGINEERING
KNOWLEDGE
FOR
MARINE ENGINEERS

by

THOMAS D. MORTON
C.Eng., F.I. Mar.E., M.I. Mech.E.
Extra First Class Engineers' Certificate

and

LESLIE JACKSON, B.Sc. (Lond.).
C.Eng., F.I. Mar.E., F.R.I.N.A.
Extra First Class Engineers' Certificate

Revised by
ANTHONY S. PRINCE
M.Ed., C. Eng., F.I. Mar.E.
Extra First Class Engineers' Certificate

THOMAS REED PUBLICATIONS
A DIVISION OF THE ABR COMPANY LIMITED

First Edition – 1975
Second Edition – 1978
Reprinted – 1982
Reprinted – 1986
Third Edition – 1994
Reprinted – 1999

ISBN 0 901281 10 7

© Thomas Reed Publications

THOMAS REED PUBLICATIONS
The Barn
Ford Farm
Bradford Leigh
Bradford-on-Avon
Wiltshire BA15 2RP
United Kingdom

Email: tugsrus@abreed.demon.co.uk

Produced by Omega Profiles Ltd, SP11 7RW
Printed and bound in Great Britain by MPG Books Ltd, Bodmin, Cornwall

PREFACE

The object of this book is to prepare students for the Certificates of Competency of the Department of Transport in the subject of Motor Engineering Knowledge.

The text is intended to cover the ground work required for both examinations. The syllabus and principles involved are virtually the same for both examinations but questions set in the First Class require a more detailed answer.

The book is not to be considered as a close detail reference work but rather as a specific examination guide, in particular **all the sketches are intended as direct application to the examination requirements.**

The best method of study is to read carefully through each chapter, practising sketchwork, and when the principles have been mastered to attempt the few examples at the end of the chapter. Finally, the miscellaneous questions at the end of the book should be worked through. The best preparation for any examination is to work on the examples, this is difficult in the subject of Engineering Knowledge as no model answer is available, nor indeed any one text book to cover all the possible questions. As a guide it is suggested that the student finds his information first and then attempts each question in the book in turn, basing his answer on either a good descriptive sketch and writing or a description covering about $1\frac{1}{2}$ pages of A4 paper in $\frac{1}{2}$ hour.

ACKNOWLEDGEMENTS TO THIRD EDITION

I wish to acknowledge the invaluable assistance given, by the following bodies, in the revision of this book:

    ABB Turbo Systems Ltd.
    New Sulzer Diesels Ltd.
    Krupp MaK Maschinenbau GmbH.
    Dr. -Ing Geislinger & Co.
    Wartsila Diesel Group.
    The Institute of Marine Engineers.
    SCOTVEC.

I also wish to extend my thanks to my colleagues at Glasgow College of Nautical Studies for their assistance.

Anthony S. Prince, 1994.

CONTENTS

CHAPTER 1		BASIC PRINCIPLES
	1 – 28	Definitions and formulae. Fuel consumption and efficiency, performance curves, heat balance. Ideal cycles, air standard efficiency, Otto, Diesel, dual, Joule and Carnot cycles. Gas turbine circuits. Actual cycles and indicator diagrams, variations from ideal, typical practical diagrams. Typical timing diagrams.
CHAPTER 2		STRUCTURE AND TRANSMISSION
	29 – 100	Bedplate, frames, crankshaft, construction, materials and stresses, defects and deflections. Lubricating oil, choice, care and testing. Lubrication systems. Cylinders and pistons. Cylinder liner, wear, lubrication. Piston rings, manufacture, defects. Exhaust valves.
CHAPTER 3		FUEL INJECTION
	101 – 128	Definitions and principles. Pilot injection. Jerk injection. Common rail. Timed injection. Indicator diagrams. Fuel valves, mechanical, hydraulic. Fuel pumps, jerk. Fuel systems.
CHAPTER 4		SCAVENGING AND SUPERCHARGING
	129 – 158	Types of scavenging, uniflow, loop, cross. Pressure charging, turbo-charging, under piston effect, parallel, series parallel. Constant pressure operation. Pulse operation. Air cooling. Turbocharger, lubrication, cleaning, surging, breakdown.

CONTENTS (cont.)

CHAPTER 5

159 – 176

STARTING AND REVERSING
General starting details. Starting air overlap, firing interval, 2- and 4- strokes, starting air valves, direct opening and air piston operated, air distributor. General reversing details, reversing of 2-stroke engines, lost motion clutch, principles, Sulzer.
Practical systems. B. and W., Sulzer starting air and hydraulic control. Rotational direction interlock.

CHAPTER 6

177 – 194

CONTROL
Governing of diesel engines., flyweight governor, flywheel effect. Proportional and reset action. Electric governor. Load sensing, load sharing geared diesels. Bridge control. Cooling and lubricating oil control. Unattended machinery spaces.

CHAPTER 7

195 – 216

ANCILLARY SUPPLY SYSTEMS
Air compressors, two and three stage, effects of clearance, volumetric efficiency, filters, pressure relief valves, lubrication, defects, automatic drain. Air vessels. Cooling systems, distilled water, lubricating oil, additives.

CHAPTER 8

217 – 234

MEDIUM SPEED DIESELS
Couplings, fluid, flexible. Clutches. Reversible gearing systems. Exhaust valve problems. Exhaust valves. Design parameters. Typical 'V' engine. Lubrication and cooling. Future development.

CHAPTER 9

235 – 256

WASTE HEAT PLANT
Boilers, package, multi-water tube sunrod, vertical vapour, Cochran, Clarkson, Gas/water heat exchangers. Silencers. Exhaust gas heat recovery circuits, natural and forced circulation, feed heating.

CHAPTER 10

257 – 270

MISCELLANEOUS
Crankcase explosions, regulations. Explosion door. Flame trap. Oil mist detector. Gas turbines. Exhaust gas emmissions.

CONTENTS (cont.)

271 – 286   TEST QUESTIONS.

287 – 296   SPECIMEN QUESTIONS

297 – 304   INDEX

CHAPTER 1

BASIC PRINCIPLES

DEFINITIONS AND FORMULAE

Isothermal Operation (PV = constant)
An ideal reversible process at constant temperature. Follows Boyle's law, requiring heat addition during expansion and heat extraction during compression. Impractical due to requirement of very slow piston speeds.

Adiabatic Operation (PV^γ = constant)
An ideal reversible process with no heat addition or extraction. Work done is equivalent to the change of internal energy. Requires impractically high piston speeds.

Polytropic Operation (PV^n = constant)
A more nearly practical process. The value of index n usually lies between unity and gamma.

Volumetric Efficiency
A comparison between the mass of air induced per cycle and the mass of air contained in the stroke volume at standard conditions. Usually used to describe 4-stroke engines and air compressors. The general value is about 90 per cent.

Scavenge Efficiency
Similar to volumetric efficiency but used to describe 2-stroke engines where some gas may be included with the air at the start of compression. Both efficiency values are reduced by high revolutions, high ambient air temperature.

Mechanical Efficiency
A measure of the mechanical perfection of an engine. Numerically expressed as the ratio between the indicated power and the brake power.

Uniflow Scavenge

Exhaust at one end of the cylinder (top) and scavenge air entry at the other end of the cylinder (bottom) so that there is a clear flow traversing the full cylinder length, *e.g.* B and W Sulzer RTA (see Fig. 1).

Loop Scavenge

Exhaust and scavenge air entry at one end of cylinder (bottom), *e.g.* Sulzer RD RND and RL. This general classification simplifies and embraces variations of the sketch (Fig.1) in cases where air and exhaust are at different sides of the cylinder with and without crossed flow loop (cross and transverse scavenge)

Brake Thermal Efficiency

The ratio between the energy developed at the brake (output shaft) of the engine and the energy supplied.

FIG 1
COMPRESSION, EXPANSION

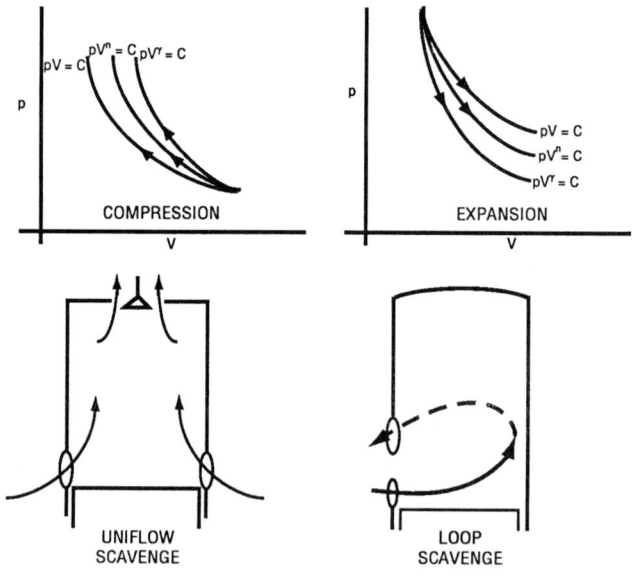

Specific Fuel Consumption
Fuel consumption per unit energy at the cylinder or output shaft, kg/kWh (or kg/kWs), 0·19 kg/kWh would be normal on a shaft energy basis for a modern engine.

Compression Ratio
Ratio of the volume of air at the start of the compression stroke to the volume of air at the end of this stroke (inner dead centre). Usual value for a compression ignition (CI) oil engine is about 12·5 to 13·5, *i.e.* clearance volume is 8 per cent of stoke volume.

Fuel – Air Ratio
Theoretical air is about 14·5 kg/kg fuel but actual air varies from about 29-44 kg/kg fuel. The percentage excess air is about 150 (36·5 kg/kg fuel).

Performance Curves Fuel Consumption and Efficiency
With main marine engines for merchant ships the optimum designed maximum thermal efficiency (and minimum specific fuel consumption) are arranged for full power conditions. In naval practice minimum specific fuel consumption is at a given percentage of full power for economical speeds but maximum speeds are occasionally required when the specific fuel consumption is much higher. For IC engines driving electrical generators it is often best to arrange peak thermal efficiency at say 70% load maximum as the engine units are probably averaging this load in operation.

The performance curves given in Fig. 2 are useful in establishing principles. The fuel consumption (kg/s) increases steadily with load. Note that halving the load does not halve the fuel consumption as certain essentials consume fuel at no load (*e.g.* heat for cooling water warming through, etc.). Willan's law is a similar illustration in steam engine practice.

Mechanical efficiency steadily increases with load as friction losses are almost constant. Thermal efficiency (brake for example) is designed in this case on the sketch for maximum at full load. Specific fuel consumption is therefore a minimum at 100% power. Fuel consumption on a brake basis increases more rapidly than indicated specific fuel consumption as load decreases due to the

**FIG 2
PERFORMANCE CURVES**

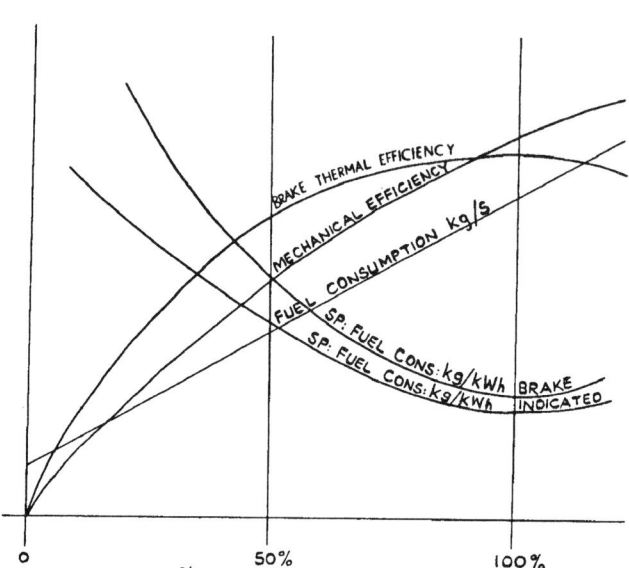

fairly constant friction loss. In designing engines for different types of duty the specific consumption minima may be at a different load point. As quoted earlier this could be about 70% for engines driving electrical generators.

Heat Balance
A simple heat balance is shown in Fig. 3.

There are some factors not considered in drawing up this balance but as a first analysis this serves to give a useful indication of the heat distribution for the IC engine. The high thermal efficiency and low fuel consumption obtained by diesel engines is superior to any other form of engine in use at present.

1. The use of a waste heat (exhaust gas) boiler gives a plant efficiency gain as this heat would otherwise be lost up the funnel.

2. Exhaust gas driven turbo-blowers contribute to high

FIG 3
SIMPLE HEAT BALANCE

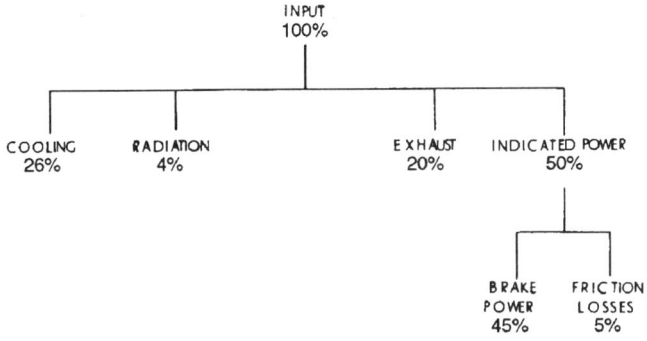

mechanical efficiency. As the air supply to the engine is not supplied with power directly from the engine, *i.e.* chain driven blowers or direct drive scavenge pumps, then more of the generated power is available for effective brake power.

Consideration of the above shows two basic flaws in the simplification of a heat balance as given in Fig. 3.

(a) The difference between indicated power and brake power is not only the power absorbed in friction. Indicated power is necessarily lost in essential drives for the engine such as camshafts, pumps, etc. which means a reduced potential for brake power.

(b) Friction results in heat generation which is dissipated in fluid cooling media, *i.e.* oil and water, and hence the cooling analysis in a heat balance should include the frictional heat effect as an assessment.

3. Cooling loss includes an element of heat energy due to generated friction.

4. Propellers do not usually have propulsive efficiencies exceeding 70% which reduces brake power according to the output power.

5. In the previous remarks no account has been taken of the increasing common practice of utilising a recovery system for heat normally lost in coolant systems.

Load Diagram.
Fig. 4 shows a typical load diagram for a slow-speed 2-stroke engine. It is a graph of brake power and shaft speed. Line 1 represents the power developed by the engine on the test bed and runs through the MCR [maximum continuous rating] point. Lines parallel to 2 represent constant values of P_{mep}. Line 3 shows the maximum shaft speed which should not be exceeded. Line 4 is important since it represents the maximum continuous power and mep, at a given speed, commensurate with an adequate supply of charge air for combustion. Line 5 represents the power absorbed by the propeller when the ship is fully loaded with a clean hull. The effect of a fouled hull is to move this line to the left as indicated by line 5a. In general a loaded vessel will operate between 4 and 5, while a vessel in ballast will operate in the region to the right of 5. The area to the left of line 4 represents overload operation.

It can be seen that the fouling of the hull, by moving line 5 to the left, decreases the margin of operation and the combination of hull fouling and heavy weather can cause the engine to become overloaded, even though engine revolutions are reduced.

IDEAL CYCLES

These cycles form the basis for reference of the actual performance of IC engines. In the cycles considered in detail all curves are frictionless adiabatic, *i.e.* isentropic. The usual assumptions are made such as constant specific heats, mass of charge unaffected by any injected fuel, etc. and hence the expression *'air standard cycle'* may be used. There are two main classifications for reciprocating IC engines, (a) spark ignition (SI) such as petrol and gas engines and, (b) compression ignition (CI) such as diesel and oil engines. Older forms of reference used terms such as light and heavy oil engines but this is not very explicit or satisfactory. Four main air standard cycles are first considered followed by a brief consideration of other such cycles less often considered. The cycles have been sketched using the usual method of P-V diagrams.

Otto (Constant Volume) Cycle
This cycle forms the basis of all SI and high speed CI engines.

The four non-flow operations combined into a cycle are shown in Fig. 5.

FIG 4
ENGINE LOAD DIAGRAM

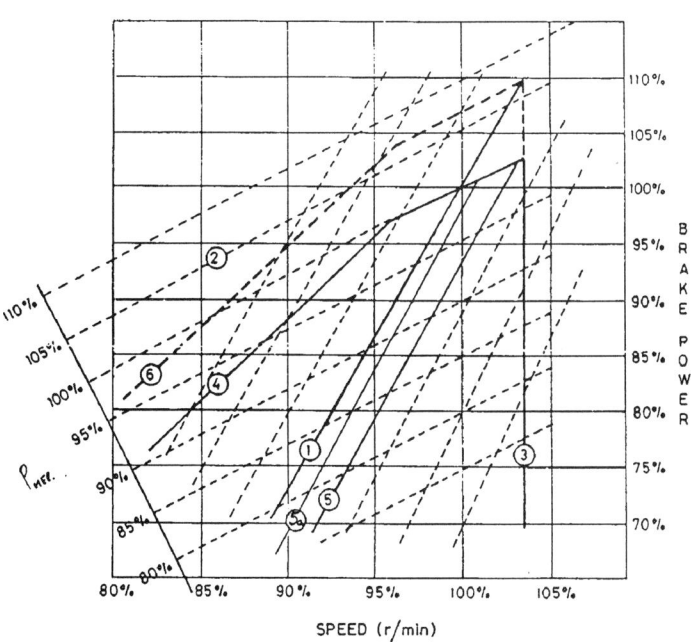

Air Standard Efficiency = Work Done/Heat Supplied
$$= \frac{\text{(Heat Supplied - Heat Rejected)}}{\text{Heat Supplied}}$$

referring to Fig. 5

Air Standard Efficiency = 1 - Heat Rejected / Heat Supplied
  = 1 - MC $(T_4 - T_1)$ /MC $(T_3 - T_2)$.
  = $1 - 1/(r^{\gamma-1})$
[using $T_2/T_1 = T_3/T_4 = r^{\gamma-1}$  where r is the compression ratio].

Note
Efficiency of the cycle increases with increase of compression ratio. This is true of the other four cycles.

FIG 5
THEORETICAL (IDEAL) CYCLES

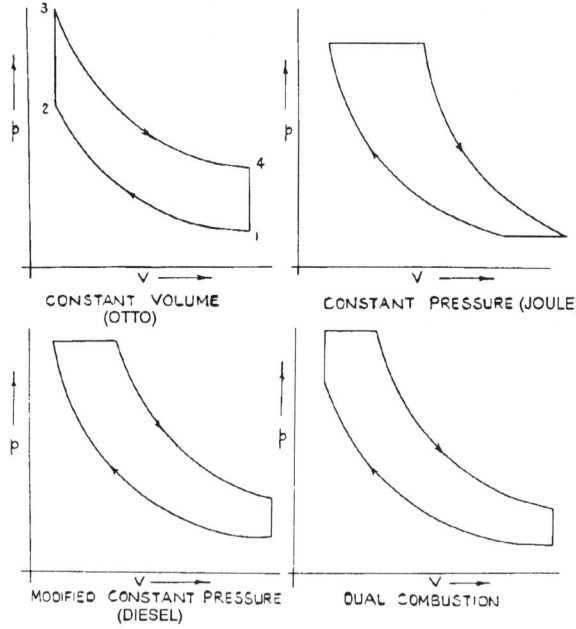

Diesel (Modified Constant Pressure) Cycle

This cycle is more applicable to older CI engines utilising long periods of constant pressure fuel injection period in conjunction with blast injection. Modern engines do not in fact aim at this cycle which in its pure form envisages very high compression ratios. The term semi-diesel was used for hot bulb engines using a compression ratio between that of the Otto and the Diesel ideal cycles. Early Doxford engines utilised a form of this principle with low compression pressures and 'hot spot' pistons. The Diesel cycle is also sketched in Fig. 5 and it may be noted that heat is received at constant pressure and rejected at constant volume.

Dual (Mixed) Cycle

This cycle is applicable to most modern CI reciprocating IC engines. Such engines employ solid injection with short fuel injection periods fairly symmetrical about the firing dead centre. The term semi-diesel was often used to describe engines working close to this cycle. In modern turbo-charged marine engines the approach is from this cycle almost to the point of the Otto cycle, *i.e.* the constant pressure period is very short. This produces very heavy firing loads but gives the necessary good combustion.

Joule (Constant Pressure) Cycle

This is the simple gas turbine flow cycle. Designs at present are mainly of the open cycle type although nuclear systems may well utilise closed cycles. The ideal cycle P-V diagram is shown in Fig 5. and again as a circuit cycle diagram on Fig. 6. in which intercoolers, heat exchangers and reheaters have been omitted for simplicity.

Other Cycles

The efficiency of a thermodynamic cycle is a maximum when the cycle is made up of reversible operations. The Carnot cycle of isothermals and adiabatics satisfies this condition and this maximum efficiency is, referring to Fig. 7 given by $(T_3 - T_1)/T_3$ where the Kelvin temperatures are maximum and minimum for the cycle. The cycle is practically not approachable as the mean effective pressure is so small and compression ratio would be excessive. All the four ideal cycles have efficiencies less than the

FIG 6
GAS TURBINE CIRCUIT-CYCLES

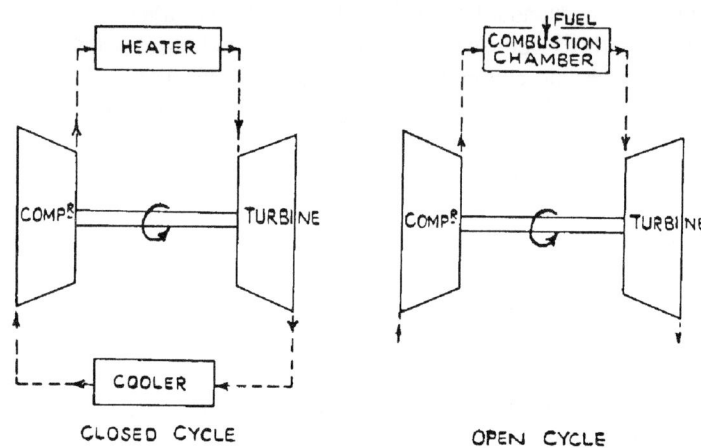

Carnot. The Stirling cycle and the Ericsson cycle have equal efficiency to the Carnot. Further research work is being carried out

FIG 7
THEORETICAL (IDEAL) CYCLES

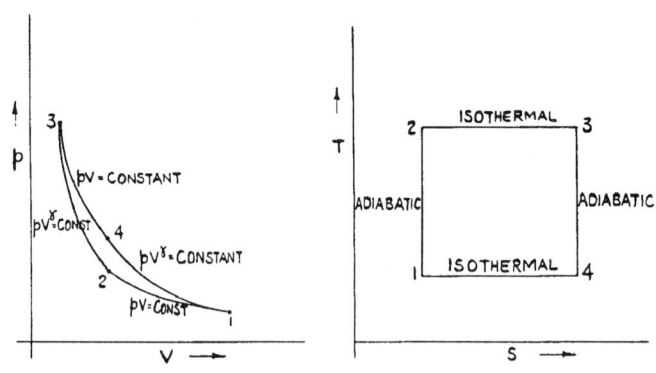

on Stirling cycle engines in an effort to utilise the high thermal efficiency potential. The Carnot cycle is sketched on both P-V and T-S axes Fig 7.

ACTUAL CYCLES AND INDICATOR DIAGRAMS

There is an analogy between the real IC engine cycle and the equivalent air standard cycle in that the P-V diagrams are similar. The differences between these cycles are now considered and for illustration purposes the sketches given are of the Otto cycle. The principles are however generally the same for most IC engine cycles.

(a) The actual compression curve (shown full line on Fig. 8.) gives a lower terminal pressure and temperature than the ideal adiabatic compression curve (shown dotted). This is caused by heat transfer taking place, variable specific heats, a reduction in γ due to gas-air mixing, etc. Resulting compression is not adiabatic and the difference in vertical height is shown as x.

(b) The actual combustion gives a lower temperature and pressure than the ideal due to dissociation of molecules caused by high temperatures. These twofold effects can be regarded as a loss of peak height of $x + y$ and a lowered expansion line below the ideal adiabatic expansion line. The loss can be regarded as clearly shown between the ideal adiabatic curve from maximum height (shown chain dotted) and the curve with initial point $x + y$ lower (shown dotted).

(c) In fact the expansion is also not adiabatic. There is some heat recovery as molecule re-combination occurs but this is much less than the dissociation combustion heat loss in practical effect. The expansion is also much removed from adiabatic because of heat transfer taking place and variation of specific heats for the hot gas products of combustion. The actual expansion line is shown as a full line on Fig. 8.

In general the assumptions made at the beginning of the section on ideal cycles are worth repeating, *i.e.* isentropic, negligible fuel charge mass, constant specific heats, etc. plus the comments above such as for example on dissociation. Consideration of these factors plus practical details such as rounding of corners due to non-instantaneous valve operation, etc. mean that the actual diagram appears as shown in the lower sketch of Fig. 8.

FIG 8
ACTUAL CYCLES (OTTO BASIS)

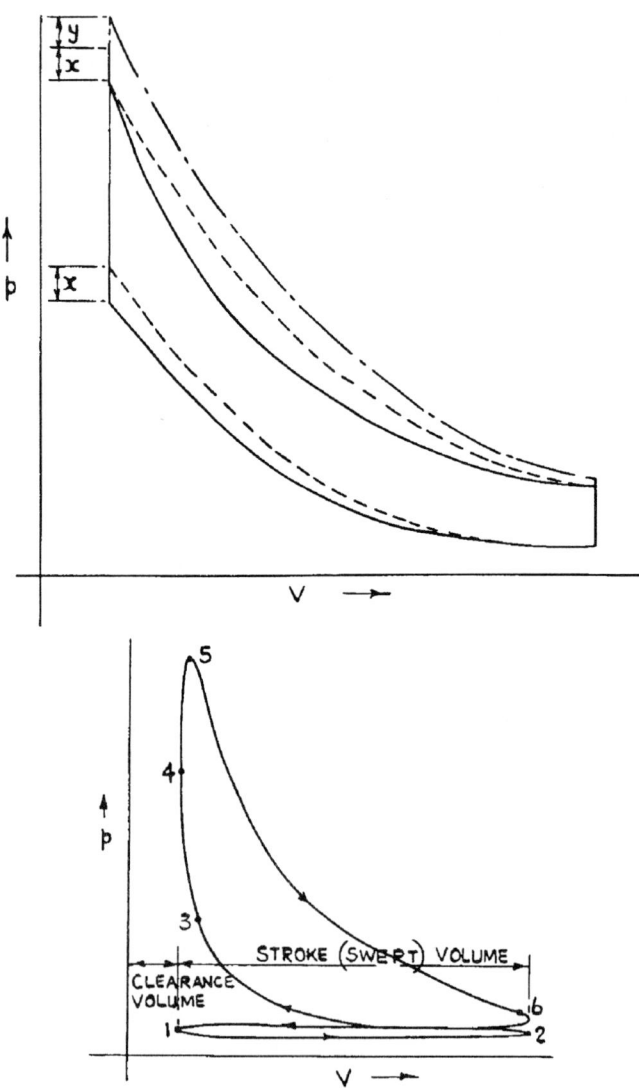

FIG 9

TYPICAL INDICATOR (POWER & DRAW) DIAGRAMS

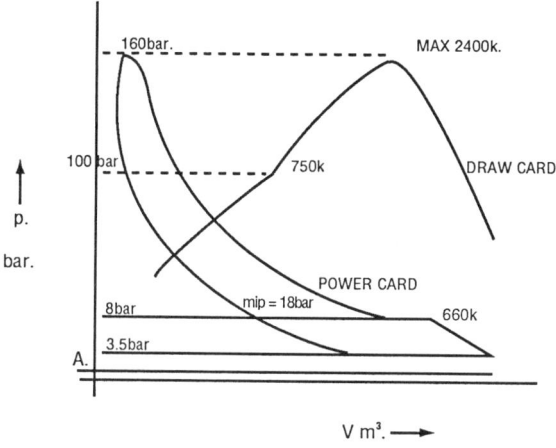

2 Stroke Cycle (CI)

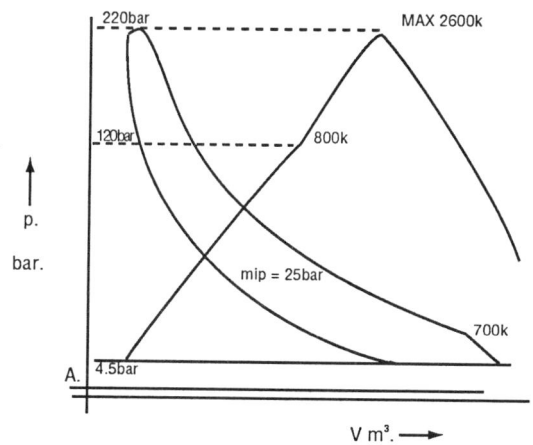

4 Stroke Cycle (CI)

Typical Indicator Diagrams

The power and draw cards are given on Fig. 9. and should be closely studied. Diagrams given are for compression ignition engines of the 2- and 4-stroke types.

Pressures and temperatures are shown on the sketches where appropriate. The draw card is an extended scale picture of the combustion process. In early marine practice the indicator card was drawn by hand–hence the name. In modern practice an 'out of phase' (90 degrees) cam would be provided adjacent to the general indicator cam. Incorrect combustion details show readily on the draw card. There is no real marked difference between the diagrams for 2-stroke or 4-stroke. In general the compression point on the draw card is more difficult to detect on the 2-stroke as the line is fairly continuous. There is no induction – exhaust loop for the 4-stroke as the spring used in the indicator is too strong to discriminate on a pressure difference of say $1/3$ bar only.

Compression diagrams are given also in Fig. 10; with the fuel shut off expansion and compression should appear as one line. Errors would be due to a time lag in the drive or a faulty indicator cam setting or relative phase difference between camshaft and crankshaft. Normally such diagrams would only be necessary on initial engine trials unless loss of compression or cam shift on the engine was suspected.

Fig. 11. is given to show the light spring diagrams for CI engines of the 2- and 4-stroke types. These diagrams are particularly useful in modern practice to give information about the exhaust – scavenge (induction) processes as so many engines utilise turbo-charge. The turbo-charge effect is shown in each case and it will be observed that there is a general lifting up of the diagram due to the higher pressures.

OTHER RELATED DETAILS

Fuel valve lift cards are very useful to obtain characteristics of injectors when the engine is running. A diagram is given in Fig 12 relating to a Doxford engine.

Typical diagram faults are normally best considered in the

BASIC PRINCIPLES 15

FIG 10
COMPRESSION DIAGRAMS

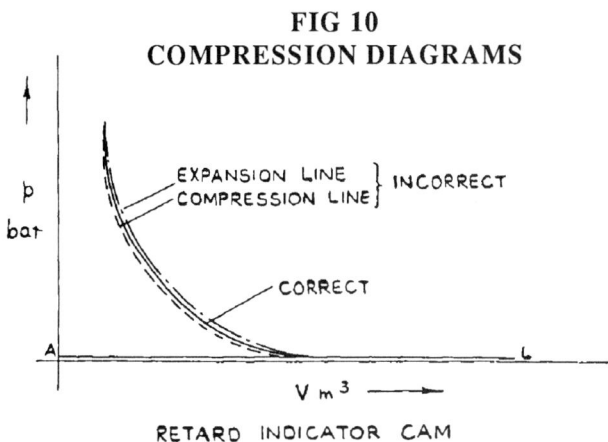

RETARD INDICATOR CAM

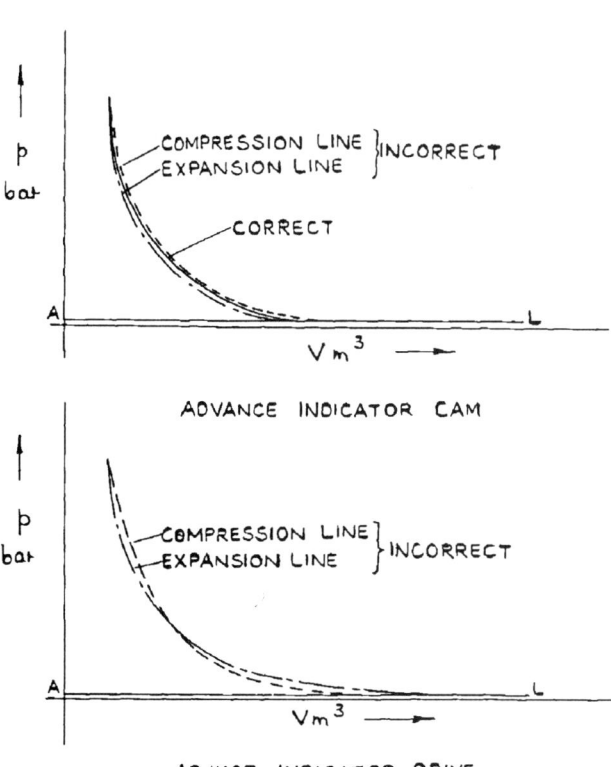

ADVANCE INDICATOR CAM

ADJUST INDICATOR DRIVE

FIG 11
TYPICAL INDICATOR (LIGHT SPRING) DIAGRAMS

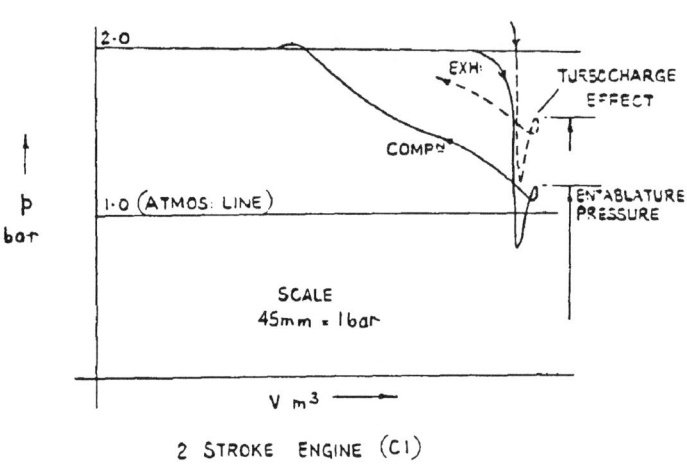

2 STROKE ENGINE (CI)

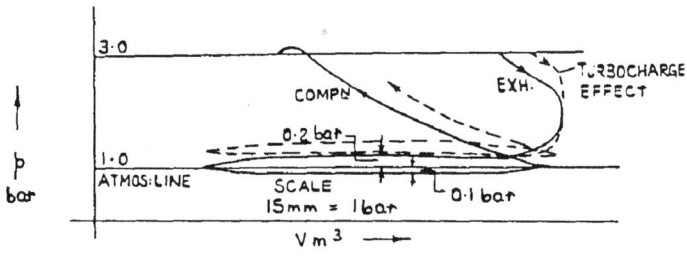

4 STROKE ENGINE (CI)

FIG 12
FUEL VALVE LIFT DIAGRAMS

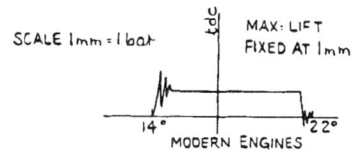

RELATED DETAILS

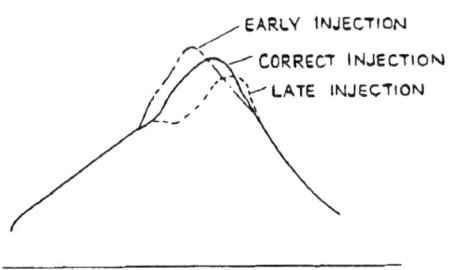

TYPICAL FAULTS SHOWN ON DRAW CARD

particular area of study where they are likely to occur. However as an introduction, two typical combustion faults are illustrated on the draw card of Fig. 12. Turbo-charge effects are also shown in Fig. 11. and compression card defects in Fig. 10. It should perhaps be stated that before attempting to analyse possible engine faults it is essential to ensure that the indicator itself and the drive are free from any defect.

Compression ratio has been discussed previously and with SI engines the limits are pre-ignition and detonation. Pinking and its relation to Octane number are important factors as are anti-knock additives such as lead tetra-ethyl Pb $(C_2H_5)_4$. Factors more specific to CI engines are ignition quality, Diesel knock and Cetane number, etc. In general these factors plus the important related topics of combustion and the testing and use of lubricants and fuels should be particularly well understood and reference should be made to the appropriate chapter in Volume 8.

Accuracy of indicator diagram calculations is perhaps worthy of specific comment. The area of the power card is quite small and planimeter errors are therefore significant. Multiplication by high spring factors makes errors in evaluation of m.i.p. also significant and certainly of the order of at least ±4%. Further application of engine constants gives indicated power calculations having similar errors. Provided the rather inaccurate nature of the final results is appreciated then the real value of the diagrams can be established. From the power card viewpoint comparison is probably the vital factor and indicator diagrams allow this. However modern practice would perhaps favour maximum pressure readings, equal fuel quantities, uniform exhaust temperature, etc. for cylinder power balance and torsionmeter for engine power. The draw card is particularly useful for compression– combustion fault diagnosis and the light spring diagram for the analysis of scavenge – exhaust considerations.

Turbo-charging
This is considered in detail later in this book but one or two specific comments relating to timing diagrams can be made now. Exhaust requires to be much earlier to drop exhaust pressure quickly before air entry and also requires to be of a longer period to allow discharge of the greater gas mass. Air period is usually slightly greater. This could mean for example in the 2-stroke cycle exhaust from 76 degrees before bottom dead centre to 56 degrees after (unsymmetrical by 20 degrees) and scavenge 40 degrees before and after. For the 4-stroke cycle air open as much as 75 degrees before top centre for 290 degrees and exhaust open 45 degrees before bottom centre for 280 degrees, i.e. considerable overlap.

Actual Timing Diagrams
Fig. 13. shows examples of actual timing diagrams for four types of engine. It will be seen that in the case of the poppet valve type of engine that the exhaust opens at a point significantly earlier than on the loop scavenged design. This is because the exhaust valve can be controlled, independently of the piston, to open and close at the optimum position. This means that opening can be carried out earlier to effectively utilise the pulse energy of the exhaust gas in the turbo-charger. The closing position can also be chosen to minimise the loss of charge air to the exhaust. With the loop scavenged engine, however, the piston controls the flow of gas into the exhaust with the result that the opening and closing of these

FIG 13 a
CRANK TIMING DIAGRAM FOR 2-STROKE LOOP SCAVENGED TURBO-CHARGED ENGINE.

EXHAUST & SCAVENGE SYMMETRICAL ABOUT BDC.

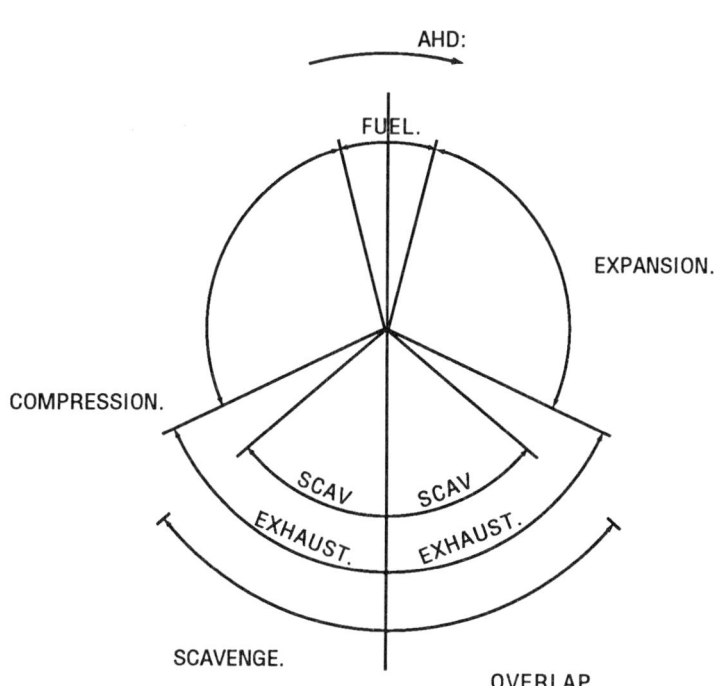

FIG 13 b
4-STROKE NATURALLY ASPIRATED ENGINE.

NOTE THE DIFFERENCE OF OVERLAP BETWEEN TURBO-CHARGED & NATURALLY ASPIRATED 4 STROKE ENGINE.

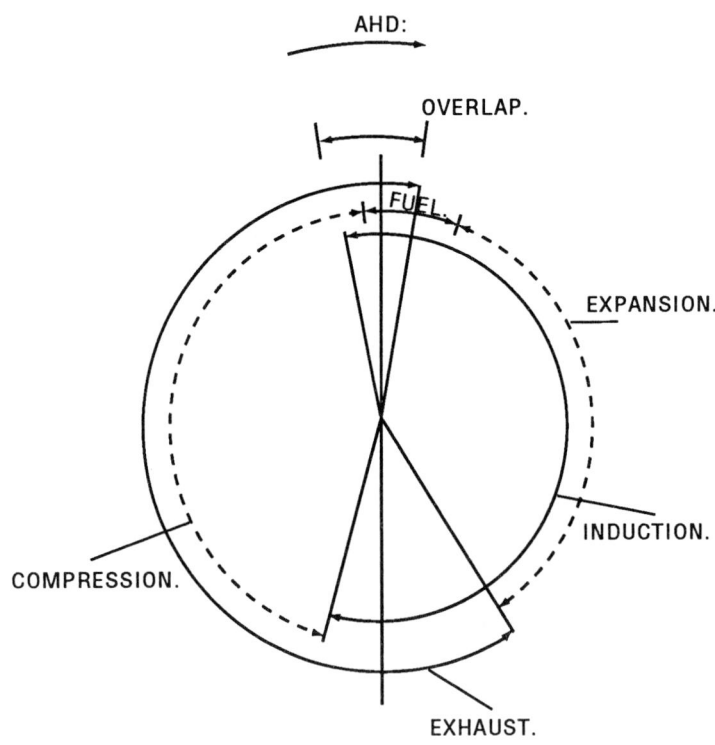

BASIC PRINCIPLES 21

FIG 13 c
4 STROKE TURBOCHARGED ENGINE

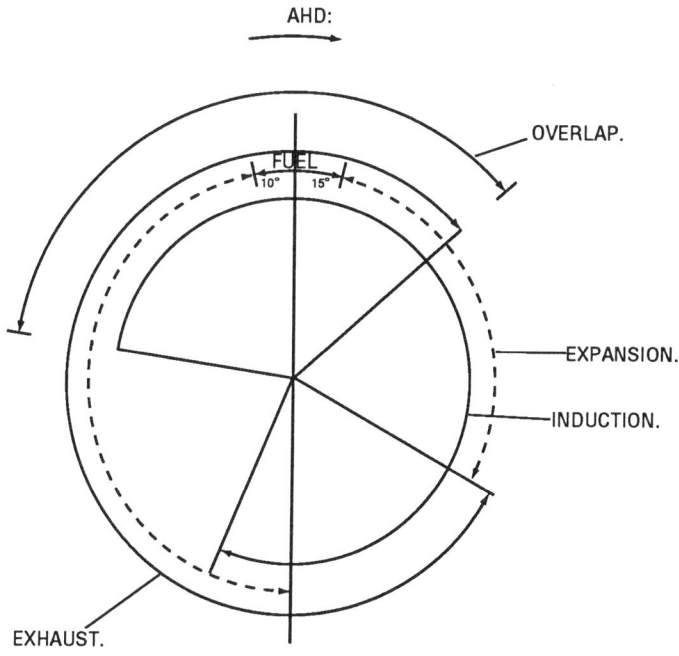

**FIG 13 d
CRANK TIMING DIAGRAM FOR 2 STROKE
TURBO-CHARGED ENGINE.
(UNI-FLOW SCAVENGE. EXHAUST
CONTROLLED BY EXHAUST V/V IN CYLINDER
COVER).**

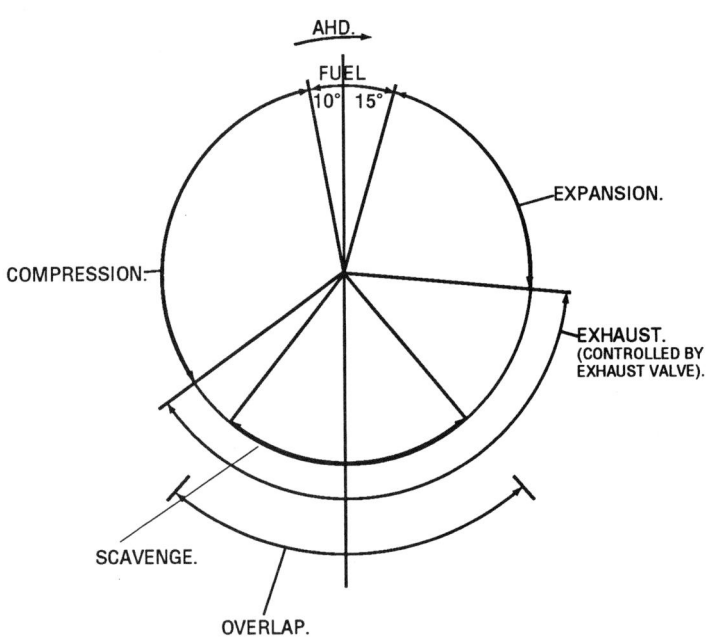

BASIC PRINCIPLES

ports are symmetrical about bottom centre. To minimise the losses of charge air to exhaust the choice of exhaust opening position is dictated by the most effective point of exhaust port closure.

Comparison of the crank timing diagrams of the naturally aspirated and turbo-charge 4-stroke diesel engine show the large degree of valve overlap on the latter. This overlap together with turbo-charging allows more efficient scavenging of combustion gases from the cylinder. The greater flow of air through the turbo-charged engine also cools the internal components and supplies a larger mass of charge air into the cylinder prior to compression commencing.

Types of Indicating Equipment

Conventional indicator gear is fairly well known from practice and manufacturers descriptive literature is readily available for precise details. For high speed engines an indicator of the 'Farnboro' type is often used. Maximum and compression pressures can be taken readily using a peak pressure indicator as sketched in Fig. 14.

**FIG 14
PRESSURE INDICATOR**

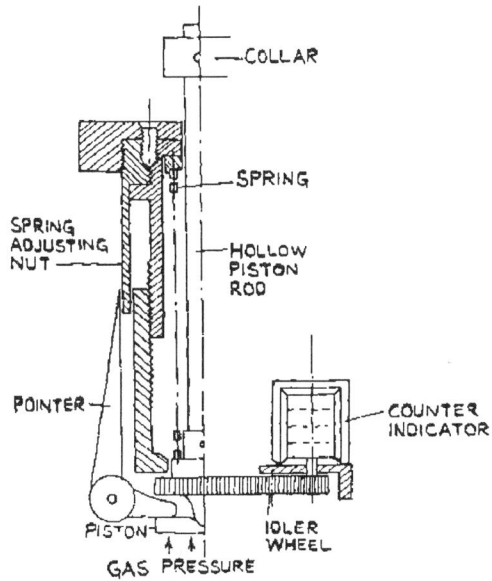

Counter and adjuster nut are first adjusted so marks on the body coincide at a given pressure on the counter with idler wheel removed. Idler wheel is now replaced. When connected to indicator cock of the engine the adjusting nut is rotated until vibrations of the pointer are damped out. Spring force and gas pressure are now in equilibrium and pressure can be read off directly on the indicating counter (Driven by toothed wheels).

Electronic Indicating

The limitations of mechanical indicating equipment have become increasingly apparent in recent years as engine powers have risen. With outputs reaching 5500 hp/cylinder inaccuracies of ±4·0% will lead to large variations in indicated power and therefore attempts to balance the engine power by this method will have only limited success. The inaccuracies stem from friction and inertia of mechanical indicator gear and errors in measuring the height of the power card.

Modern practice utilises electronic equipment to monitor and analyse the cylinder peak pressures and piston position and display onto a VDU [video display unit]. The cylinder pressure is measured by a transducer attached to the indicator cock. Engine position is detected by a magnetic pick up in close proximity to a toothed flywheel. The information is fed to a microprocessor, where it is averaged over a number of engine cycles, before calculations are made as to indicated power and mean effective pressure. Fig. 15. The advantages of this type of equipment is that:

1. It supplies dynamic operational information.

This means that injection timing can be measured while the engine is running. This is a more accurate method of checking injection timing since it allows for crankshaft twist while the engine is under load, unlike static methods which do not.

2. Can compare operating conditions with optimum performance.

This should lead to improvements in fuel economy and thermal efficiency.

3. Can produce a load diagram for the engine, clearly defining the safe operating zone for the engine.

4. Can produce trace of fuel pressure rise in fuel high pressure lines. Valuable information when diagnosing fuel faults.

Operational experience with this type of equipment has pointed

FIG 15
ELECTRONIC INDICATOR EQUIPMENT

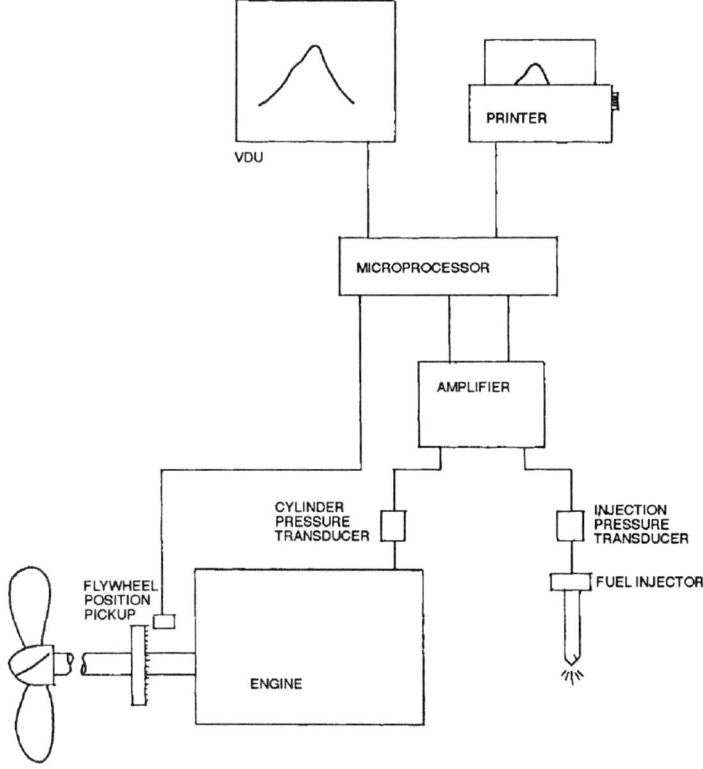

to unreliability of the pressure transducers when connected continuously to the engine. To overcome this problem manufacturers are experimenting with alternative methods of measuring cylinder pressure. One alternative is to permanently attach a strain gauge to one cylinder head stud of each cylinder. Since the strain measured is a function of cylinder pressure this information can be fed to the microprocessor. The increased reliability of this technique will allow the equipment to be permanently installed allowing power readings to be taken at any time. This type of equipment can be used to measure many other engine parameters to aid diagnosis and accurately monitor performance such as fuel pump pressure etc.

Fatigue

Fatigue is a phenomenon which affects materials that are subjected to cyclic or alternating stresses. Designers will ensure that the stress of a component is below the yield point of the material as measured on the familiar stress/strain graph. However if that component is subjected to cyclic stresses it may fail at a lower value due to fatigue. The most common method of displaying information on fatigue is the S~N curve Fig. 16. This information is obtained from fatigue tests usually carried out on a Wohler machine in which a standard specimen is subjected to an alternating stress due to rotation. The specimen is tested at a particular stress level until failure occurs. The number of cycles to failure is plotted against stress amplitude on the S~N curve. Other specimens are tested at different levels of stress. When sufficient data have been gathered a complete curve for a particular material may be presented.

It can be seen from Fig. 16. that, in the case of ferrous materials, there is a point known as the "fatigue limit". Components stressed below this level can withstand a infinite number of stress reversals without failure. Since:

$$\text{Stress} = \frac{\text{load}}{\text{CSA}}$$

It can be seen that reducing the stress level on a component involves increasing the CSA [cross sectional area] resulting in a weight penalty. In marine practice the weight implications are, in general, secondary to reliability and long-life and so components are usually stressed below the fatigue limit. This is not the case in,

BASIC PRINCIPLES 27

FIG 16
WOHLER MACHINE FOR ZERO MEAN STRESS FATIGUE TESTING

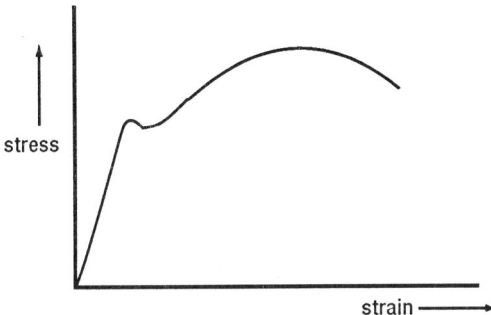

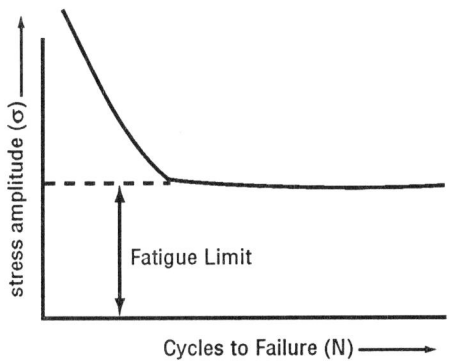

Typical S/N curve FOR FERROUS MATERIAL

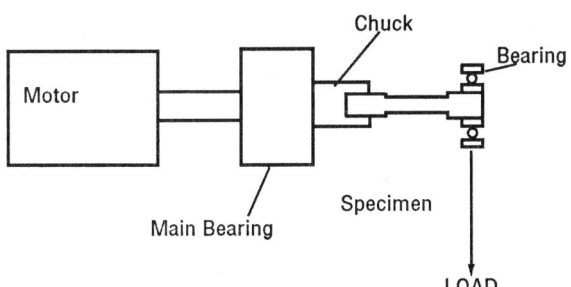

for example, aeronautical practice where weight is a major consideration. In this situation the component designer would compromise between weight and stress levels and from the S~N curve would calculate, with the addition of a safety margin, the number of cycles the component could withstand before failure occurs. The working life of 4-stroke medium speed diesel bottom end bolts are calculated in this way.

CHAPTER 2

STRUCTURE AND TRANSMISSION

The engine structure consisting of the bedplate and "A" frames or, in more modern designs the frame section must fulfil the following fundamental requirements and properties.

Strength – is necessary since considerable forces can be exerted. These may be due to out of balance effects, vibrations, gas force transmission and gravitational forces.

Rigidity – is required to maintain correct alignment of the engine running gear. However, a certain degree of flexibility will prevent high stresses that could be caused by slight misalignment.

Lightness – is important, it may enable the power weight ratio to be increased. Less material would be used, bringing about a saving in cost. Both are important selling points as they would give increased cargo capacity.

Toughness – in a material is a measure of its resilience and strength, this property is required to enable the material to withstand the fatigue conditions which prevail.

Simple design – if manufacture and installation are simplified then a saving in cost will be realised.

Access – ease of access to the engine transmission system for inspection and maintenance, and in the first instance installation, is a fundamental requirement.

Dimensions – ideally these should be as small as possible to keep engine containment to a minimum in order to give more engine room space.

Seal – the transmission system container must seal off effectively the oil and vapours from the engine room.

Manufacture – Modern engines increasingly are manufactured in larger modular sections that allow for convenience in assembly.

BEDPLATE

This is a structure that may be made of cast iron, prefabricated steel, cast steel, or a hybrid arrangement of cast steel and prefabricated steel.

Cast iron one piece structures are generally confined to the smaller engines. That is, medium speed engines rather than the larger slow speed cross-head type of engine. This is due to the problems that arise as the size of the casting increases. These problems include poor flow of material to the extremities of the mould, poor grain size control which leads to a lack of homogeneity of strength and soundness and poor impurity segregation. In addition to these problems cast iron has poor performance in tension and its modulus of elasticity is only half that of steel hence for the same strength and stiffness a cast iron bedplate will require to be manufactured from more material. This results in weight penalty for larger cast iron bedplates when compared with a fabricated bedplate of similar dimensions. Cast iron does, however, enjoy certain advantages for the construction of smaller medium and high speed engines. Castings do not require heat treatment, cast iron is easily machined, it is good in compression, the master mould can be re-used many times which results in reduced manufacturing costs for a series of engines. The noise and vibration damping qualities of cast iron are superior to that of fabricated steel. As outputs increase nodular cast iron, due to its higher strength, is becoming more common for the manufacture of medium speed diesel engine bedplates.

Modern cast iron bedplates for medium speed engines are generally, but not exclusively, a deep inverted "U" shape which affords maximum rigidity for accurate crankshaft alignment. The crankcase doors and relief valves are incorporated within this structure. In this design the crankshaft is "underslung" and the crankcase closed with a light unstressed oil tray. Fig. 17.

As outputs of medium speed engines increase some manufacturers choose the alternative design in which the crankcase and bedplate are separate components. The crankshaft being "embedded" in the bedplate. Fig. 18.

FIGURE 17
SECTION THROUGH ENGINE BLOCK OF MEDIUM SPEED ENGINE WITH UNDERSLUNG CRANKSHAFT.

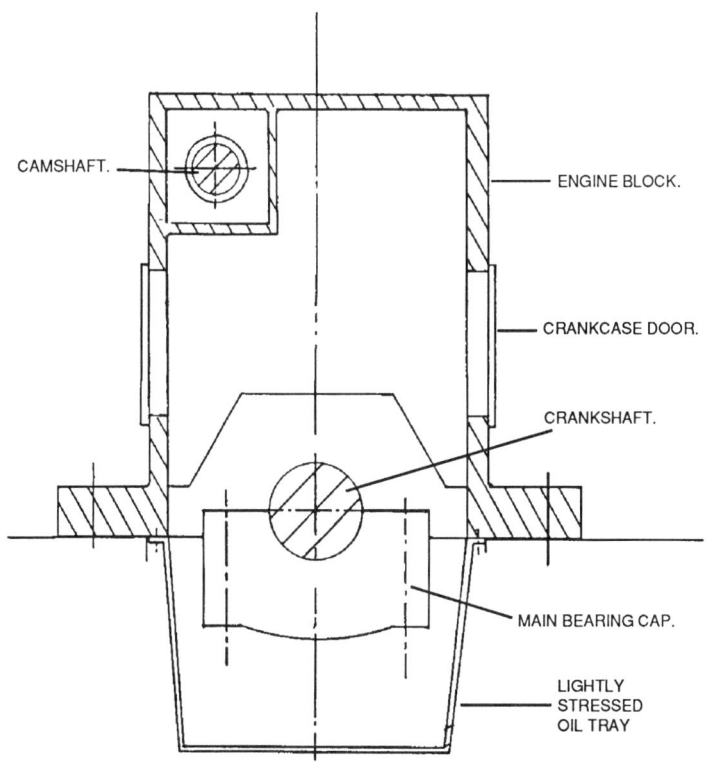

FIGURE 18
MEDIUM SPEED ENGINE BEDPLATE WITH EMBEDDED CRANKSHAFT

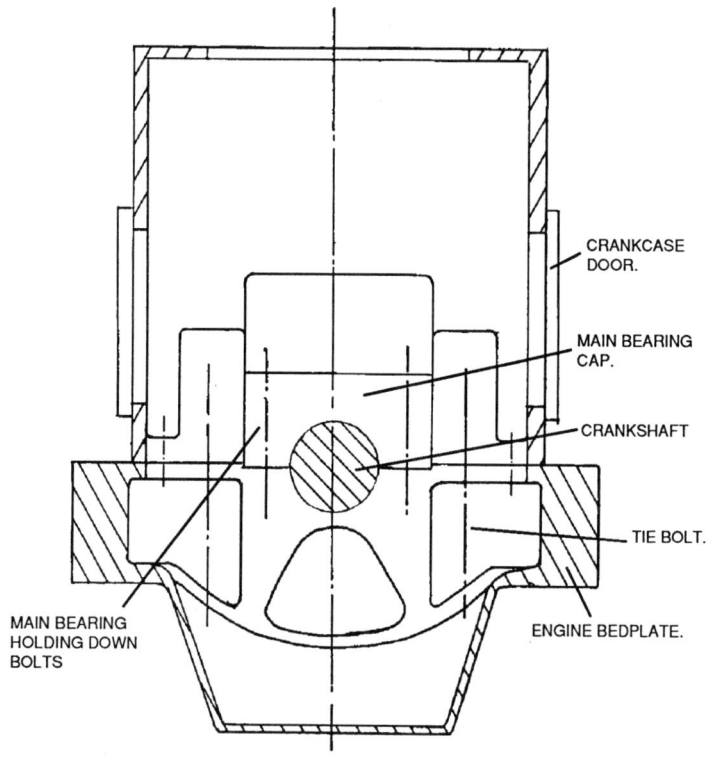

When welding techniques and methods of inspection improved and larger furnaces became available for annealing, the switch to prefabricated steel structure with its saving in weight and cost was made. It must be remembered that the modulus of elasticity for steel is nearly twice that of cast iron, hence for similar stiffness of structure roughly half the amount of material would be required when using steel.

Early designs were entirely fabricated from mild steel but radial cracking due to cyclic bending stress imposed by the firing loads was experienced on the transverse members in way of the main bearings. The adoption of cast steel, with its greater fatigue strength, for transverse members has eliminated this cracking. Modern large engine bedplates are constructed from a combination of fabricated steel and cast steel. Modern designs consist of a single walled structure fabricated from steel plate with transverse sections incorporating the cast steel bearing saddles attached by welding Fig. 19. To increase the torsional, longitudinal and lateral rigidity of the structure suitable webbing is incorporated into the fabrication.

It is modern practice to cut the steel plate using automatic contour flame cutting equipment. Careful preparation is essential prior to the welding operation:
- Since it is necessary to prepare the edges of the cut plate it is necessary to make an allowance for this when cutting.
- Equipment is set correctly to ensure smallest heat affected zone, [HAZ].
- Welding consumables stored and used correctly to prevent hydrogen contamination of HAZ which could lead to post anneal hydrogen cracking.

Following the welding operation the welds are inspected for surface cracking and sub-surface flaws. The surface inspection is carried out by the dye-penetrant method or the magnetic particle method while the sub-surface flaws are inspected by ultrasonic testing. Flaws in welds would be cut out, rewelded and tested.

The bedplate is then stress relieved by heating the whole structure to below the Lower Critical Temperature of the material in a furnace and allowing it to cool slowly over a period of days. When cool the structure is shot blasted and the welds again tested before the bedplate is machined.

FIGURE 19
MODERN FABRICATED SINGLE WALLED BEDPLATE WITH CAST STEEL BEARING SADDLE.

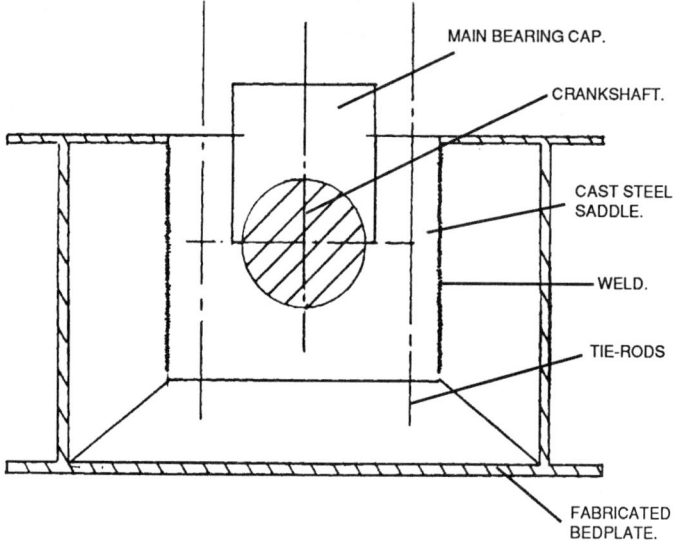

In order to minimise stresses due to bending in the bedplate, without a commensurate increase in material, tie-rods are used to transmit the combustion forces. Two tie-rods are fitted to each transverse member and pass, in tubes, through the entire structure of the engine from bedplate to cylinder cooling jacket. They are pre-stressed at assembly so that the engine structure is under compression at all times. Engines utilising the opposed piston principle have the combustion loads absorbed by the running gear and do not require to be fitted with tie-rods. To minimise bending tie-rods are placed as close as possible to to the shaft centre line.

Fig. 20 shows diagrammatically the arrangement used in the Sulzer engine. By employing jack bolts, under compression, to

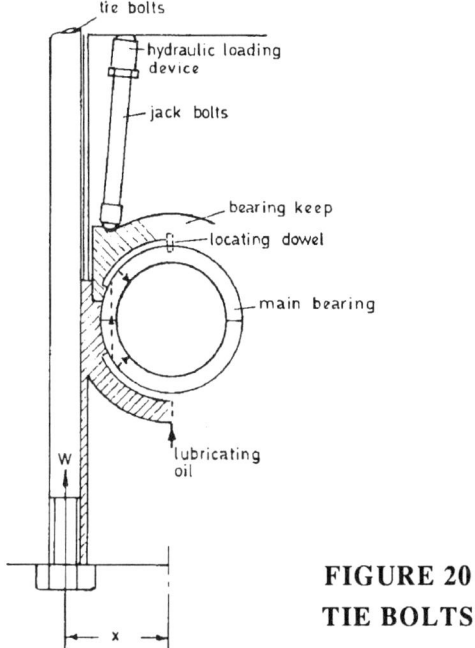

**FIGURE 20
TIE BOLTS**

retain the bearing keeps in position the distance x is kept to a minimum. Hence the bending moment Wx where W is the load in the bolt, is also a minimum.

Because of their great length, tie-rods in large slow speed diesel engines may be in two parts to facilitate removal. They are also liable to vibrate laterally unless they are restrained. This usually takes the form of pinch bolts that prevent any lateral movement.

Although tie-rods are tightened, to their correct pretension during assembly, they should be checked at intervals. This is accomplished by:
- Connecting both pre-tensioning jacks to two tie-rods lying opposite each other. Fig. 21.
- Operating the pump until the correct hydraulic pressure is reached. This pressure is maintained.
- Checking the clearance between the nut and intermediate ring with a feeler gauge. If any clearance exists then the nut is tightened onto the intermediate ring and the pressure

FIGURE 21

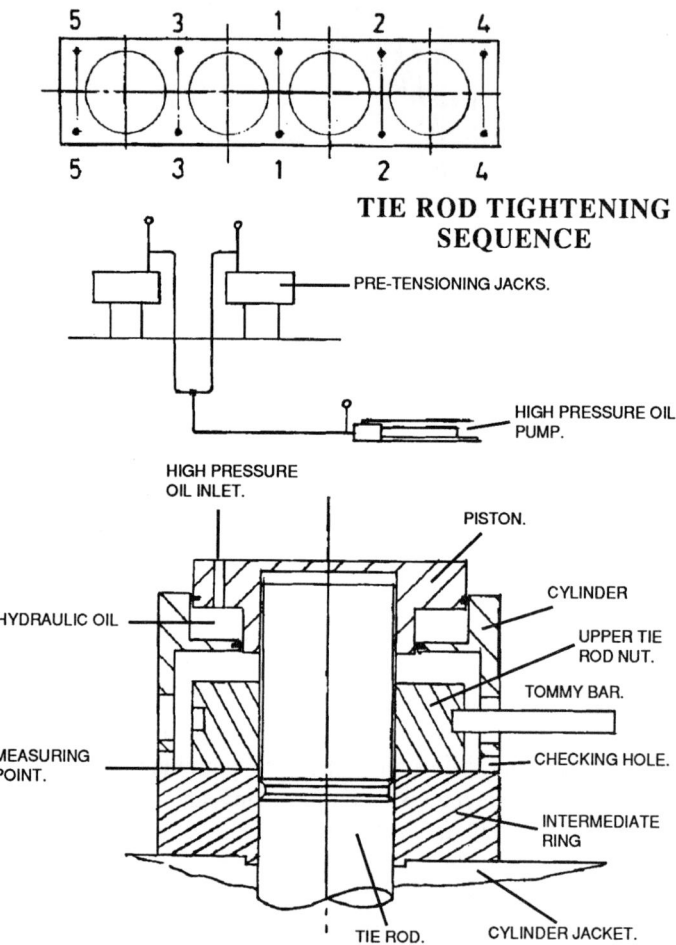

TIE ROD TIGHTENING SEQUENCE

PRE-TENSION JACK

FIGURE 22
GENERAL ARRANGEMENT OF ENGINE STRUCTURE

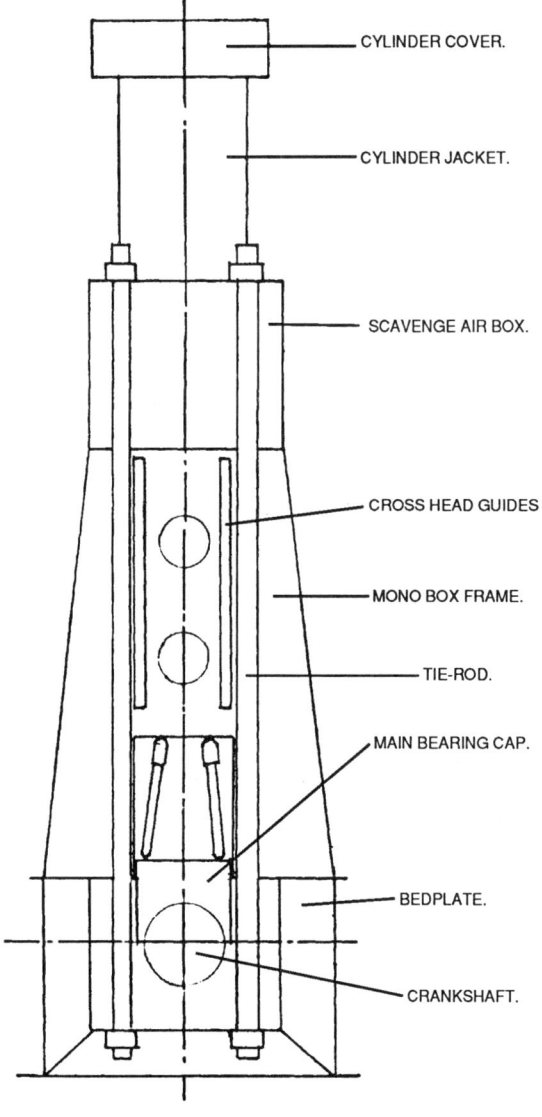

released. If no clearance is found the pressure can be released and the hydraulic jacks removed.

When using hydraulic tensioning equipment it is essential that it is maintained in good order and the accuracy of the pressure gauges are checked regularly.

If when inspecting the engine it is found that a tie-rod has broken then it must be immediately replaced. If the breakage that occurs is such that the lower portion is short and can be removed through the crankcase, the upper part can be withdrawn with relative ease from the top. If, however, the breakage leaves a long lower portion it is necessary to cut the rod to be removed in sections through the crankcase.

"A" Frames or Columns

The advent of the long and super-longstroke slow-speed diesel engines has resulted in an increase in lateral forces on the guide. This is due to the use of relatively short connecting-rods to reduce the overall height of this type of engine which results in an increased angle and a higher lateral force component Fig. 23.

In order to maintain structural rigidity under these conditions designers tend not to utilise the traditional "A" frame arrangement, preferring instead the "monoblock" structure which consists of a continuous longitudinal beam incorporating the crosshead guides Fig. 24.

The advantages of the monoblock design are:
- Greater structural rigidity.
- More accurate alignment of crosshead.
- Forces are distributed throughout the structure resulting in a lighter construction.
- Improved oil tightness.

The construction of monoblock structures is similar to that described for bedplates above.

Holding down arrangements

The engine must be securely attached to the ship's structure in such a way as to maintain the alignment of the crankshaft within the engine structure.

There are two main methods of holding the engine to the ship's structure.

STRUCTURE AND TRANSMISSION 39

**FIG 23
INCREASED CONNECTING ROD ANGLE GIVING HIGHER LATERAL FORCES.**

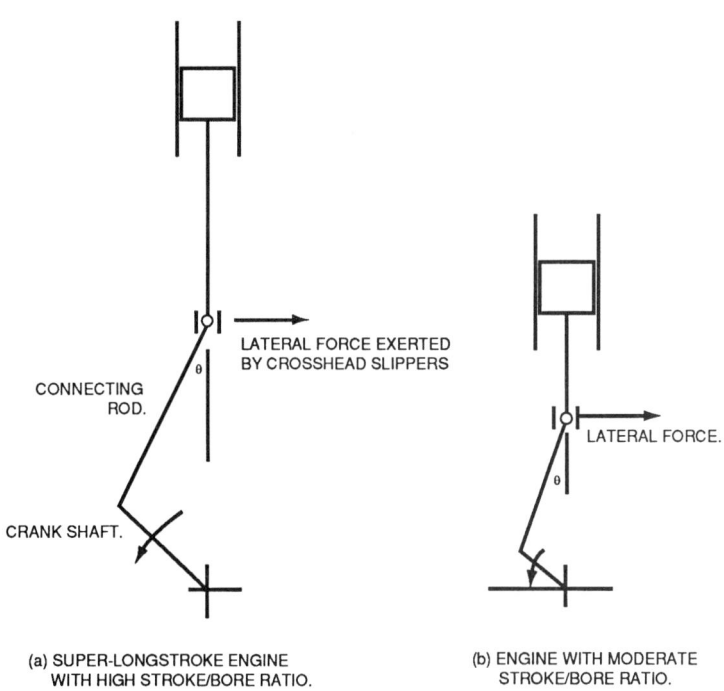

(a) SUPER-LONGSTROKE ENGINE WITH HIGH STROKE/BORE RATIO.

(b) ENGINE WITH MODERATE STROKE/BORE RATIO.

1. By rigid foundations onto the ship's structure.
2. Mounting the engine onto the ship's structure via resilient mountings.

Rigid foundations
In this method, the most common, fitted chocks are installed between the engine bedplate and the engine seating on the tank-top. The holding down bolts passing through the chocks. During installation of the engine great care must be taken to ensure that

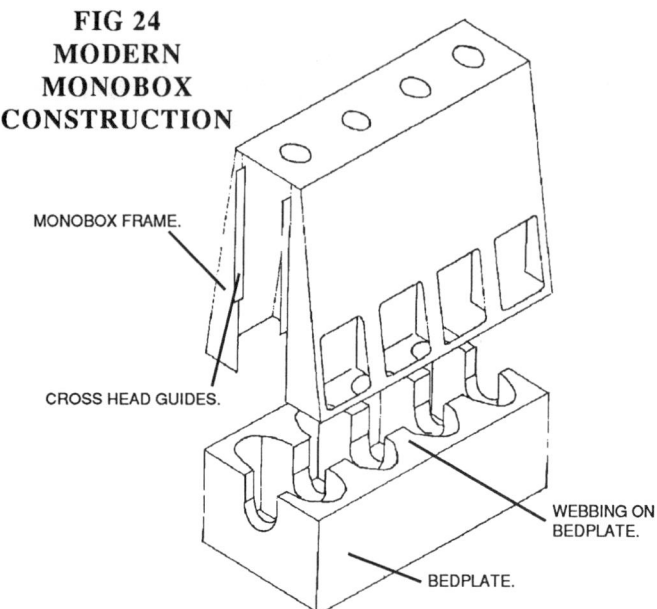

FIG 24 MODERN MONOBOX CONSTRUCTION

MONOBOX FRAME.
CROSS HEAD GUIDES.
WEBBING ON BEDPLATE.
BEDPLATE.

there is no distortion of the bedplate which would lead to crankshaft misalignment. In addition, great care must be taken to correctly align the crankshaft to the propeller shaft. The engine is initially installed on jacking bolts which are adjusted to establish its correct location in relation to the propeller shaft. When the engine is correctly positioned, and crankshaft deflections indicate no misalignment, the space between the bedplate and seating is measured and chocks are manufactured. To facilitate the fitting of chocks the top-plate of the engine seating is machined with a slight outboard facing taper. The chocks, usually made from cast iron, are individually fitted and must bear load over at least 85% of their area. The surface of the bedplate and the underside of the top plate that will make contact with the holding down bolt and nut faces are machined parallel to ensure that no bending stresses are transferred to the bolt. As the holding down bolts and chocks are installed the jacking bolts are removed.

Holding down bolts for modern slow-speed installations tend to be the long sleeved type and are hydraulically tensioned Fig. 25. This type of bolt, because of its greater length, has greater elasticity and is therefore less prone to cracking than the superceded short unsleeved bolt. The bolts are installed through the

FIG 25
LONG SLEEVED HOLDING DOWN BOLT

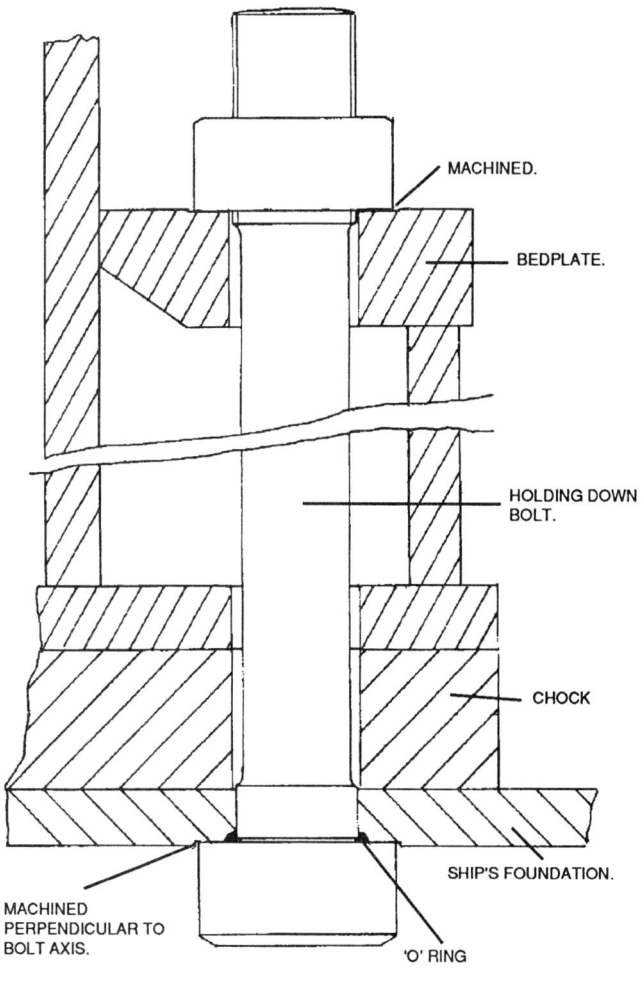

top plate and a waterproof seal is usually effected with "O" rings. Fitted bolts are installed adjacent to the engine thrust.

The holding down bolts should only withstand tensile stresses and should not be subjected to shear stresses. The lateral and transverse location is maintained by side and end chocking. The number of side chocks depends upon the length of the engine Fig 26.

It is extremely important that the engine is properly installed during building. The consequences of poor initial installation are

FIG 26
SIDE AND END CHOCKING

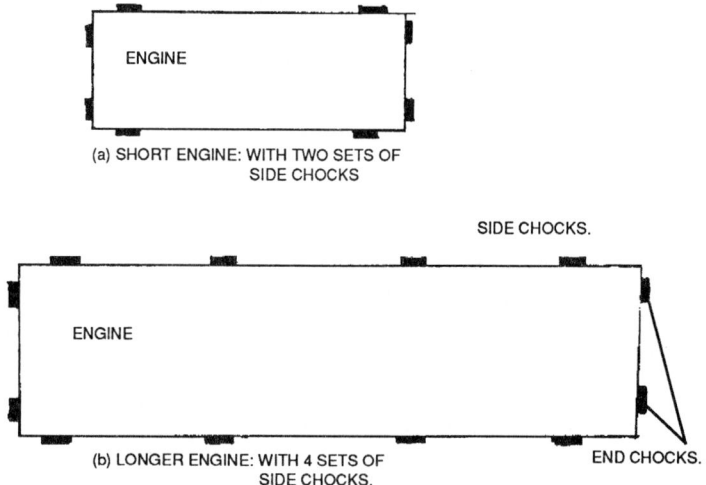

(a) SHORT ENGINE: WITH TWO SETS OF SIDE CHOCKS

SIDE CHOCKS.

(b) LONGER ENGINE: WITH 4 SETS OF SIDE CHOCKS.

END CHOCKS.

extremely serious since it may lead to fretting of chocks, the foundation and bedplate, slackening and breakage of holding down bolts and ultimately in a worsening in the alignment of the engine. To maintain engine alignment it is important to inspect the bolts for correct tension and the chocks for evidence of fretting and looseness.

An alternative to the traditional chocking materials of cast iron or steel is epoxy resin. This material, originally used as an adhesive and protective coating, was developed as a repair technique to enable engines to be realigned without the need for the machining of engine seatings and bedplate. It is claimed that the time taken to accomplish such a repair is reduced so reducing the overall costs. Although initially developed as a repair technique the use of epoxy resin chocks is becoming widespread for new buildings.

Resin chocks do not require machined foundation surfaces thus reducing the preparation time during fabrication. The engine must be correctly aligned with the propeller shaft without any bedplate distortion. This is done in the usual way with the exception that it is set high by about $1/1000$ of the chock thickness to allow for very slight chock compression when the installation is bolted down. The tank top and bedplate seating surfaces must then be thoroughly cleaned with an appropriate solvent to remove all traces of paint, scale and oil.

Because resin chocks are poured it is necessary for "dams", made from foam strip, to be set to contain the liquid resin. Plugs or the holding down bolts are now inserted. Fitted bolts being sprayed with a releasing agent, ordinary bolts being coated with a silicone grease to prevent the resin from adhering to the metal. The outer sides of the chocks are now dammed with thin section plate, fashioned as a funnel to facilitate pouring and 15 mm higher than the bedplate to give a slight head to the resin. This is also coated to prevent adhesion. Prior to mixing and pouring of the resin it is prudent to again check the engine alignment and crankshaft deflections.

The resin and activator are mixed thoroughly with equipment that does not entrain air. The resin is poured directly into the dammed off sections. Curing will take place in about 18 hrs if the temperature of the chocking area is maintained at about 20 to

25°C. The curing time can be up to 48 hours if the temperatures are substantially below this. During the chocking operation it is necessary to take a sample of resin material from each batch for testing purposes.

The advantages claimed for "pourable" epoxy resin chocks over metal chocks include:
- Quicker and cheaper installation.
- Lower bolt tension by a factor of 4 when compared to metal chocks.
- Elimination of misalignment due to fretting and bolt slackening. Because of the intimate fit of resin chocks and the high coefficient of friction between resin and steel the thrust forces are distributed to all chocks and bolts thus reducing the total stress on fitted bolts by about half Fig 27.

FIG 27
POURED RESIN CHOCKS

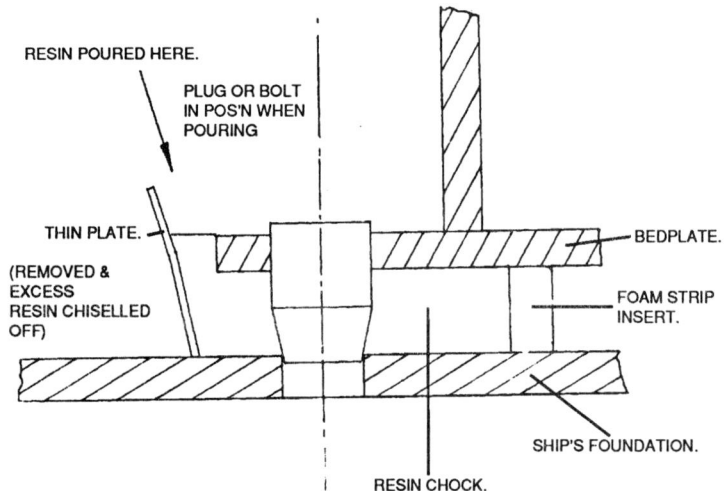

STRUCTURE AND TRANSMISSION

Resilient mountings

A possible disadvantage of rigidly mounted engines is the likelihood of noise being transmitted through the ship's structure. This is undesirable on a passenger carrying vessel where low noise and vibration levels are necessary for passenger comfort. Many manufacturers are now installing diesel engines on resilient mountings.

Diesel engines generate low frequency vibration and high frequency structure borne noise. The adoption of resilient mountings will successfully reduce both noise and vibration.

The reduction of noise and vibration of resilient and non resiliently mounted engines can be seen in Fig. 28.

**FIG 28
REDUCTION IN STRUCTURE BORNE NOISE
ACHIEVED BY RESILIENT MOUNTINGS.**

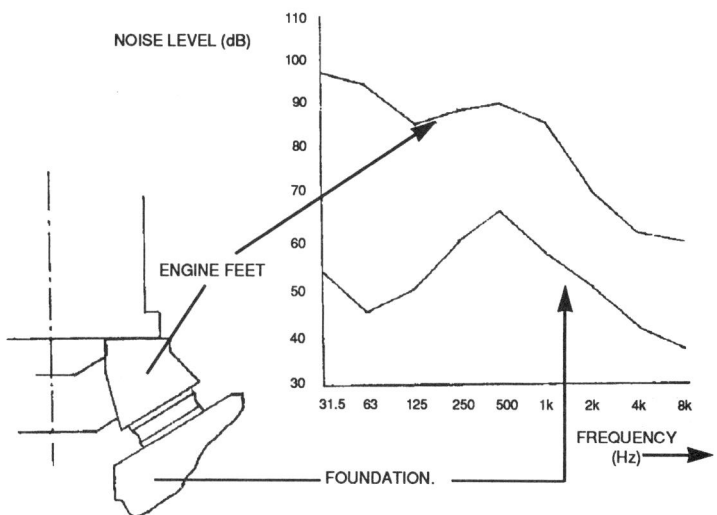

In Fig. 29 it can be seen that the diesel engine is aligned and rigidly mounted to a fabricated steel sub-frame. This can be either via solid or resin chocks. The sub-frame is then resiliently mounted to the ship's structure on standard resilient elements.

In geared engine applications the engine is again mounted, via solid or resin chocks, to a sub-frame which is resiliently mounted to the ship's structure. The engine is then coupling to the reduction gearbox through a highly elastic coupling. It is necessary to limit the amount of lateral and longitudinal movement of the engine, relative to the ship's structure. This is accomplished by stopper devices built into the holding down arrangement.

When starting and stopping resiliently mounted diesel engines large transitory amplitudes of vibration can be encountered. One manufacturer's solution to this problem is to install a hydraulic locking device. This device, shown in Fig. 30, has a piston with a connection via a shut off valve between both sides. During normal running the connecting valve is open allowing the piston to displace oil between upper and lower chamber freely. During starting and stopping this valve will be closed effectively preventing relative movement between engine and ship's structure.

Crankshafts

A crankshaft is the backbone of the diesel engine. Despite being subjected to very high complex stresses the crankshaft must none-the-less be extremely reliable since not only would the costs of failure be very high but, also, the safety of the vessel would be jeopardised

Crankshafts must be extremely reliable, if we examine the stresses to which a crankshaft is subjected then we may appreciate the need for extreme reliability.

Fig. 31. shows a crank unit with equivalent beam systems. Diagram (a) indicates the general, central variable loaded, built-in beam characteristic of a crank throw supported by two main bearings. If the bearings were flexible. *e.g.* spherical or ball, then a simply supported beam equivalent would be the overall characteristic.

Examining the crank throw in greater detail, diagram (b), shows that the crank pin itself is like a built-in beam with a distributed

FIG 29
SUB-FRAME TYPE RESILIENT ENGINE MOUNTING

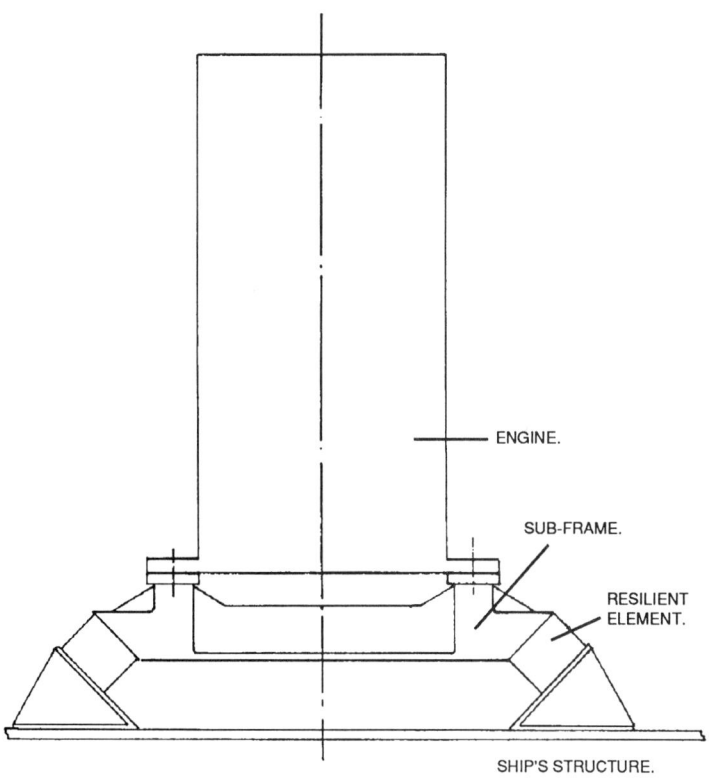

**FIG 30
HYDRAULIC LOCKING DEVICE FOR ENGINE
MOVEMENT LIMITATION DURING
STARTING/STOPPING**

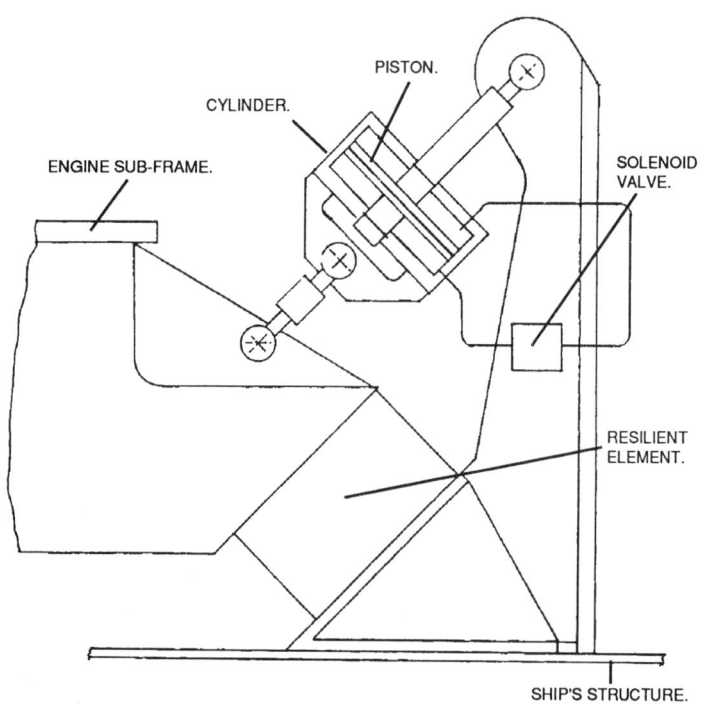

FIG 31
STRESSES IN CRANKSHAFT

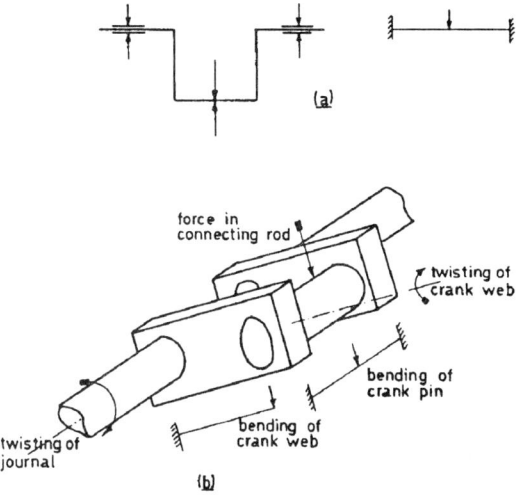

load along its length that varies with crank position. Each crank web is like a cantilever beam subjected to bending and twisting. Journals would be principally subjected to twisting, but a bending stress must also be present if we refer back to diagram (a).

Bending causes tensile, compressive and shear stresses. Twisting causes shear stress.

Because the crankshaft is subjected to complex fluctuating stresses it must resist the effects of fatigue. To this end the material and the method of manufacture must be chosen carefully. For fatigue considerations forging is preferable to casting. This is because, unlike casting, forgings exhibit directional "grain flow". The properties of the material in the direction transverse to the grain flow being significantly inferior to those in the direction longitudinal to the grain flow. Under these circumstances the drop in fatigue strength may be as must as 25 to 35% with similar reductions in strength and ductility. Forging methods, therefore, ensure that the principal direction of grain flow is parallel to the major direct stresses imposed on the crankshaft. Fig. 32.

The materials chosen for forged and cast crankshafts are essentially the same. The composition of the steel will vary

FIG 32
DIRECTION OF GRAIN FLOW IN FORGED CRANKSHAFT.

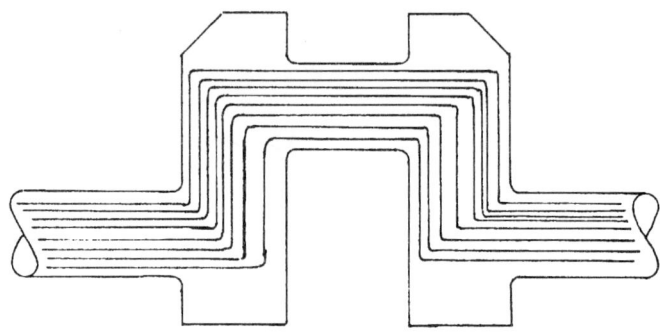

depending upon the bearing type chosen. For a crankshaft with white metal bearings a steel of 0.2% carbon may be chosen, this will have a UTS of approximately 425 to 435 MN/m^2. For higher output applications with harder bearing materials the carbon content is in the range of 0.35% to 0.4% which raises the UTS to approximately 700 MN/m^2. To increase the hardness of the shaft still further alloying agents such as chromium-molybdenum and nickel are added. For smaller engines such as automotive applications the crankshafts are surface hardened and fatigue resistance increased by nitriding.

There are two broad categories of crankshafts:
1. One piece construction.
2. "Built up" from component parts.

1. One piece construction

One piece construction, either cast or forged, is usually restricted to smaller medium and high speed engines. Following the casting or forging operation the component is rough machined to its approximate final dimensions and the oil passages are drilled. The fillet radius and crankpin are then cold rolled to improve the fatigue resistance and reduce the micro-defects on the surface.

Following machining the crankshaft is then tested for surface and sub-surfaced defects.

2. Built up crankshafts

There are 3 categories of built up crankshafts:
Fully built up; webs are shrunk onto journals and crankpins Fig 33a.
Semi-built up; webs and crankpin as one unit shrunk onto the journals Fig. 33b.
Welded construction; webs, journals and crankpin are welded together. Fig. 33c.

Fully and semi-built up construction

To minimise the risk of distortion of fully and semi built crankshafts, assembly is carried out vertically. Various jigs are required to ensure the correct crank angles and to provide support for the crankshaft. The webs are heated only to about 400°C. and the journals and pins inserted. Raising the temperature higher would bring the steel to the critical temperature and change the material's characteristics. When the assembly has cooled the web material adjacent to the journal will be in tension. The level of stress in this region must be well below the limit of proportionality to ensure that the material does not yield which would reduce the force of the web on the journal and lead to fretting and probable slippage. To ensure an adequate shrinkage an allowance of $1/550$ to $1/700$ of the shaft diameter is usual. Exceeding this allowance would simply increase the stress in the material without appreciably improving the grip.

When the component parts of the crankshaft have been built up the journals and pins are machined and the fillet radii cold rolled Fig. 34a. The crankshaft is then subjected to thorough surface and sub-surface tests using, for example, ultra-sound and metal particle techniques.

To reduce the weight and the out of balance effects of the crankshaft, the crankpins may be bored out hollow. Fig. 33.

The fully built up crankshaft has generally been superceded by the semi-built up type which display improved "grain flow" in webs and crankpin, are stiffer and can be shorter due to a reduction in the thickness of the webs.

**FIG 33
3 TYPES OF BUILT-UP CRANKSHAFT**

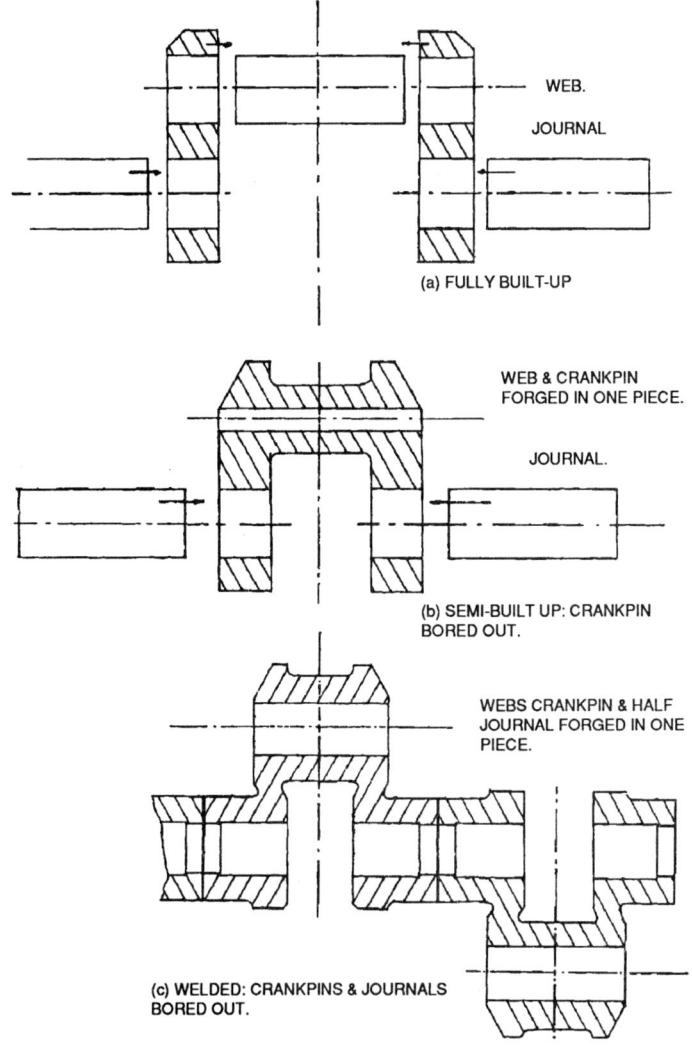

STRUCTURE AND TRANSMISSION 53

FIG 34
DETAILS OF CRANKSHAFT

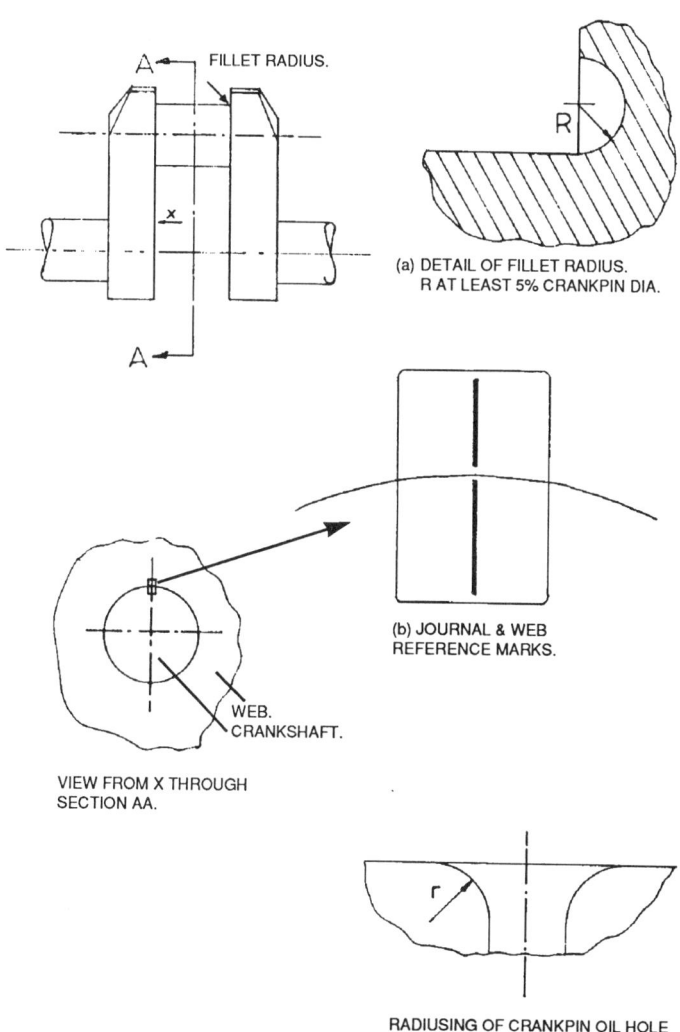

(a) DETAIL OF FILLET RADIUS.
R AT LEAST 5% CRANKPIN DIA.

(b) JOURNAL & WEB REFERENCE MARKS.

VIEW FROM X THROUGH SECTION AA.

RADIUSING OF CRANKPIN OIL HOLE
(WHERE APPLICABLE)

The effectiveness of the grip due to shrinking depends upon:
1. The shrinkage allowance. The correct allowance will result in the correct level of stress in the web and journal.
2. The quality of surface finish of the journal and web. Good quality surface finish will give the maximum contact area between web and journal.

Dowels are not used to locate the shrink since this would introduce a stress concentration which could lead to fatigue cracking.

When a crankshaft is built up by shrink fitting, reference marks are made to show the correct relative position of web and journal. Fig. 34b. These marks should be inspected during crankcase inspections. Slippage could occur:
- If starting air is applied to the cylinders when they contain water or fuel, or when the turning gear is engaged.
- If an attempt is made to start the engine when the propeller is constrained by, for example, ice or a log.
- If during operation the propeller strikes a submerged object.
- If the engine comes to a rapid unscheduled stop.

Following these circumstances a crankshaft inspection must be made and the reference marks checked. Slippage will result in the timing of the engine being altered which if not corrected will result in inefficient operation and possible poor starting. If the slippage is small, for example, up to 15° then re-timing of the affected cylinders may be considered. If, however, the slippage is such that re-timing may affect the balance of the engine then the original journal and web relative positions must be restored. This is accomplished by heating the affected web whilst cooling the journal with liquid nitrogen and jacking the crankshaft to its original position. Needless-to-say this would be accomplished by specialist personnel under controlled conditions.

Welded Construction
The development of the large marine cross head 2-stroke engine will undoubtedly result in higher outputs without an accompanying increase in physical size. These requirements impose limitations on the traditional shrink fitting of journals and webs. To transmit the torques required the traditional shrink fitting method requires that the web is of a minimum width and radial thickness. This will

FIG 35
TWO OPTIONS OF WELDED CRANKSHAFTS

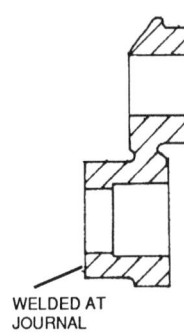

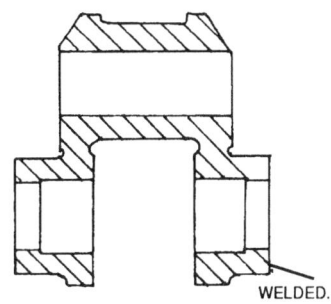

(a) HALF CRANK
WELDED AT JOURNAL &
CRANKPIN

(b) CRANKTHROW FORGED COMPLETE.
WELDED AT JOURNAL.

inevitably lead to a larger crankshaft and consequently a larger engine.

Welded construction is seen as a viable solution to this problem and one major manufacturer has invested considerable resources in developing such a construction.

There are two methods of assembly:

1. Welding two crankarms together then making a crankshaft by welding the crankarms together Fig. 35a.

2. Forging a crankthrow complete with half journals then welding them with others to form the crankshaft Fig. 35b.

The welding technique chosen is submerged arc narrow gap. This technique is automated and produces a relatively small heat affected zone [HAZ], which produces minimal residual stresses and distortion.

At the completion of welding the crankshaft is heated in a furnace to 580°C followed by a slow cooling period. Following heat treatment the crankshaft is tested using ultrasonic and metal particle techniques. If flaws are found the weld is machined out and rewelded.

The advantages claimed for welded crankshafts are:
1. Reduced principle dimensions of the engine.
2. Reduced web thickness results in a considerable reduction in weight.
3. Reduced web thickness allows journal lengths to be increased resulting in lower specific bearing loads.
4. Freedom to choose large bearing diameters without overlap restrictions.
5. Increased stiffness of crankshaft resulting in higher natural frequencies of torsional vibration.

Crankshaft defects and their causes

Misalignment
If we assume that alignment was correct at initial assembly then possible reasons for misalignment are as follows:
1. Worn main bearings. Caused by incorrect bearing adjustment leading to overloading. Broken, badly connected or choked lubricating oil supply pipes causing lubrication starvation. Contaminated lubricating oil. Vibration forces.
2. Excessive bending of engine framework. This could be caused by incorrect cargo distribution but is unlikely, more probable that the cause would be grounding of the vessel, it being re-floated in a damaged condition. It is essential that all bearing clearances be checked and crankshaft deflections taken after such an accident.

Vibration
This can be caused by: incorrect power balance, prolonged running at or near critical speeds, slipped crank webs on journals, light ship conditions leading to impulsive forces from the propeller (e.g. forcing frequency four times the revs. for a four-bladed propeller), the near presence of running machinery, excessive wear down of the propeller shaft bearing (this in bad weather conditions can lead to whipping of the shafting).

Vibration accentuated stresses, they can be increased to exceed fatigue limits and considerable damage could result. It can lead to things working loose, *e.g* coupling bolts, bearing bolts, bolts securing balance masses to crank webs and lubricating oil pipes.

Other causes

Incorrect manufacture leading to defects is fortunately a rare occurrence. In the past, failure has been caused by: slag inclusions, heat treatment and machining defects, for example badly radiused oil holes and fillets. Careless use of tools resulting in impact marks on crankpins and journals can also lead to failure. These defects all result in the the creation of stress concentrations which, because of cyclic nature of the loading of the crankshaft can raise the local level of stress in the component above the level of the fatigue limit on the S ~N graph, Fig. 16. Chapter 1, resulting in fatigue cracking and ultimate failure. This can be exacerbated if the engine is run at or close to the critical speed. The critical speed of an engine is the crankshaft speed which causes the crankshaft to vibrate at its natural frequency of torsional vibration. In other words it is the speed which induces resonance. The consequence of resonance is to cause the crankshaft to vibrate in the torsional mode with large amplitudes. Stress, being proportional to amplitude, increases and may rise sufficiently to reduce the number of working cycles of the crankshaft before failure occurs.

Bottom end bolts on medium and high speed 4-stroke diesel engines are subjected to fluctuating cyclic stresses and are therefore also exposed to potential fatigue failure. 4-stroke engine bottom end bolts experience large fluctuations of stress during the cycle. This is due to the inertia forces experienced in reversing the direction of the piston over top dead centre on the exhaust stroke. The forces experience by bottom end bolts in this situation is high. Reference to the S ~N graph in chapter 1 will show that to ensure maximum serviceability, stresses should be commensurate with a level below the fatigue limit. Since :

$$\text{stress} = \frac{\text{load}}{\text{area}}$$

it can be seen that for a given load the stress can only be reduced by increasing the area and therefore increasing the size and weight of the bottom end bolt. Designers opt for a compromise, they design a bolt that will experience a level of stress ABOVE that of the fatigue limit and specify the number of cycles the bolt should remain in service before it is replaced. It is therefore of vital importance that the running hours of 4-stroke engines are known in order to monitor the safe working life of bottom end bolts.

In addition to this, designers will specify that bottom end bolts:
- Are manufactured to high standards of surface finish.
- Have rolled threads.
- Be of the "waisted" design with generous radii.
- Have increased diameter at mid shank to reduce vibration.
- Be tightened accurately to the required level.

During maintenance bolts should be examined for mechanical damage which would cause a stress concentration. Damaged bolts should be replaced.

Fretting Corrosion
Occurs where two surfaces forming part of a machine, which in theory constitute a single unit, undergo slight oscillatory motion of a microscopic nature.

It is believed that the small relative motion causes removal of metal and protective oxide film. The removed metal combines with oxygen to form a metal oxide powder that may be harder than the metal (certainly in the case of ferrous metals) thus increasing the wear. Removed oxide film would be repeatedly replaced, increasing further the amount of damage being done.

Fretting damage increased with load, amplitude of movement and frequency. Hardness of the metal also effects the attack, in general damage to ferrous surfaces is found to decrease as hardness increases.

Oxygen availability also contributes to the attack, if oxygen level is low the metal oxides formed may be softer than the parent metal thus minimising the damage. Moisture tends to decrease the attack.

Bearing Corrosion
In the event of fuel oil and lubricating oil combining in the crankcase, weak acids may be released which can lead to corrosion of copper lead bearings. The lead is removed from the bearing surface so that the shaft runs on nearly pure copper, this raises bearing temperature so that lead rises to the surface and is removed. The process is repeated until failure of the bearing takes place. Scoring of crankshaft pins can then occur. Use of detergent types of lubricating oil can prevent or minimise this type of corrosion. The additives used in the oil to give it detergent properties would be alkaline, in order to neutralise the weak acids.

Water in the lubricating oil can lead to white metal attack and the formation of a very hard black incrustation of tin oxide. This oxide may cause damage to the journal or crankpin surface by grinding action.

Bearing Clearances and Shaft Misalignment

Bearing clearances can be checked in a variety of ways, a rough check is to observe the discharge of oil, in the warm condition, from the ends of the bearings. Feeler gauges can be used, but for some of the bearings they can be difficult to manoeuvre into position in order to obtain readings. Clock (or as they are sometimes called, dial) gauges can be very effective and accurate providing the necessary relative movement can be achieved, this can prove to be very difficult in the larger types of engine. Finally, the use of lead wire necessitating the removal of the bearing keeps.

Main bearing clearances, should be zero at the bottom. If they are not, then the crankshaft is out of alignment. Some engines are provided with facilities for obtaining the bottom clearance (if any) of the main bearings with the aid of special feelers, without the need to remove the bearing keep. Another method is to first arrange in the vertical position a clock gauge so that it can record the movement of the crank web adjacent to the main bearing. The main bearing keep is then removed, shims are withdrawn and the keep is replaced and tightened down. The vertical movement of the shaft, if any, is observed on the dial gauge.

Obviously, if the main bearing clearance is not zero at the bottom the adjacent bearing or bearings are high by comparison and the shaft is out of alignment.

Crankshaft alignment can be checked by taking deflections. If a crank throw supported on two main bearings is considered, the vertical deflection of the throw in mid span is dependent upon: shaft diameter, distance between the main bearings, type of main bearing, and the central load due to the running gear. A clock gauge arranged horizontally between the crank webs opposite the crank pin and ideally at the circumference of the main journal (see Fig. 36.) will give a horizontal deflection, when the crank is rotated through one revolution, that is directly proportional to the vertical deflection.

In Fig. 36(a). it is assumed that main bearings are in correct alignment and no central load is acting due to running gear, then

FIG 36
CHECKING CRANKSHAFT ALIGNMENT

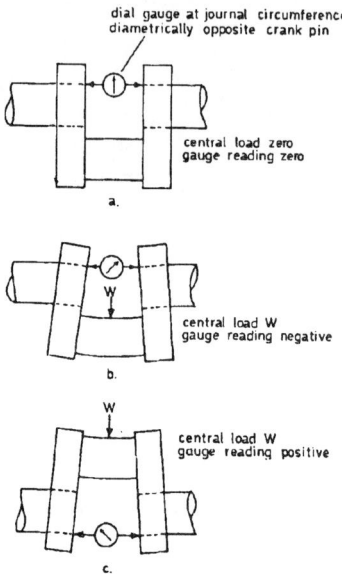

vertical deflection of the shaft would be small – say zero. With running gear in place and crank at about bottom centre the webs would close in on the gauge as shown – this is negative deflection. With crank on top centre webs open on the gauge – this is positive deflection.

In practice the gauge must always be set up in the same position between the webs each time, otherwise widely different readings will be obtained for similar conditions. An alternative is to make a proportional allowance based on distance from crankshaft centre. Obviously the greater the distance from the crankshaft centre the greater will be the difference in gauge readings between bottom and top centre positions.

Since, due to the connecting rod, it is generally not possible to have the gauge diametrically opposite crank pin centre when the crank is on bottom centre an average of two readings would be taken, one either side during the turning of the crank.

The following table shows some possible results from a six cylinder diesel engine:

GAUGE READINGS IN mm/100

CRANK POSITION	CYLINDER NUMBER					
	1	2	3	4	5	6
x	0	0	0	0	0	0
p	5	2	6	-8	-3	1
t	10	3	12	-14	-8	4
s	5	3	6	-8	-6	3
y	-2	2	-2	0	0	-2
b=(x+y)/2	-1	1	-1	0	0	-1
Vertical mis-alignment (t-b)	11	2	13	-14	-8	5
Horizontal mis-alignment (p-s)	0	-1	0	0	3	-2

The dial gauge would be set at zero when crank is in, say, port side near bottom position and gauge readings would be taken at port horizontal, top centre, starboard horizontal and starboard side near bottom positions. Say x, p, t, s and y as per Fig. 37., but before taking each reading the turning gear should be reversed to unload the gear teeth, otherwise misleading readings may be obtained.

Engines with spherical main bearings will have greater allowances for crankshaft misalignment than those without. Spherical bearings are used when increased flexibility is required for the crankshaft. This would be the case for opposed piston engines with large distances between the main bearings, so instead of having a built-in beam effect the arrangement is more likened to a simply supported beam, with its larger central deflection for a given load.

From the vertical misalignment figures and by referring to Fig.37. the reader should be able to deduce that, the end main bearing adjacent to No. 1 cylinder and the main bearing between Nos. 3 and 4 cylinders are high.

Vertical and horizontal misalignment can be checked against the permissible values supplied by the engine builder, often in the form of a graph as per Fig. 38. If any values exceed or equal maximum permissible values then bearings will have to be adjusted or renewed where required. Indication of incorrect bearing

FIG. 37
CRANK POSITIONS FOR DEFLECTION

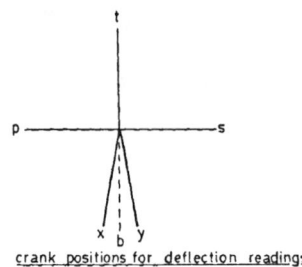

crank positions for deflection readings

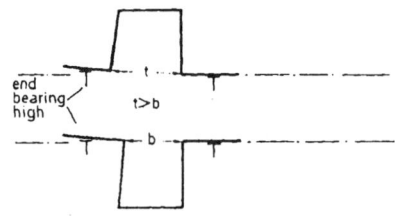

effect of bearing misalignment

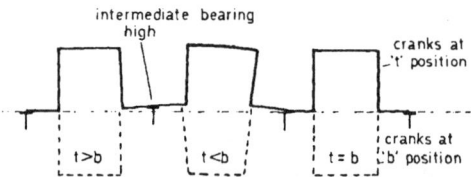

clearances may be given when the engine is running. In the case of medium or high speed diesels, load reversal at the bearings generally occurs. With excessive bearing clearances loud knocking takes place, white metal then usually gets hammered out.

If bearing lubrication for a unit is from the same source as piston cooling, then a decrease in the amount of cooling oil return, may be observed in the sight glass, together with an increase in its temperature.

If bearing clearances are too small, overheating and possible seizure may take place. Oil mist and vapour at a particular unit

FIG. 38
MAXIMUM VALUES

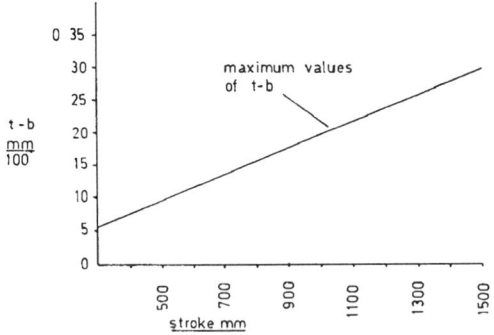

may be observed to increase – together with a hot bearing, this may lead to a crankcase explosion.

Regular checks must be made to ascertain the oxidation rate of the oil. If this is increasing then high temperatures are being encountered. n.b. as the oil oxidises (burns) its colour blackens.

CHOICE, MAINTENANCE AND TESTING OF LUBRICATING OIL FOR MAIN CIRCULATING SYSTEM

Choice

If the engine is a 'trunk type' then fuel and deleterious deposits from its combustion products may find their way into the crankcase. The oil should therefore be one which has detergent properties, these oils are sometimes called 'Heavy Duty'. Additives in these oils deter the formation of deposits by keeping substances, such as carbon particles, in suspension. They also counteract the corrosive effect of sulphur compounds, some of the fuels used may be low in sulphur content, in this case the alkaline additive in the lubricating oil could be less.

Detergent oils may not be able to be water washed in a centrifuge, it is always advisable to consult the supplier.

Straight mineral oil, generally with an anti-oxidant and corrosion inhibitor added, is the type normally used in diesels whose working cylinder is separate from the crankcase.

Maintenance

When the engine is new correct pre-commissioning should give a clean system free from sand, metal, dust, water and other foreign matter. To clear the system of contaminants all parts must be vibrated by hammering or some other such method to loosen rust flakes, scale and weld spatter (if this is not done then these things will work loose when the engine is running and cause damage). A good flushing oil should then be used and clear discharges obtained from pipes before they are connected up, filters must be opened up and cleaned during this stage. Finally, the flushing operation should be frequently repeated with a new charge of oil of the type to be used in the engine.

When the engine is running, continuous filtration and centrifugal purification is essential.

Oxidation of the oil is one of the major causes of its deterioration, it is caused by high temperatures. This may be due to:

1. Small bearing clearances (hence insufficient cooling).

2. Not continuing to circulate the oil upon stopping the engine. In the case of oil cooled piston types, piston temperatures could rise and the static oil within them become overheated.

3. Incorrect use of oil preheater for the purifier, *e.g* shutting off oil before the heat or running the unit part full.

4. Metal particles of iron and copper can act as catalysts that assist in accelerating oxidation action. Rust and varnish products can behave in a similar fashion.

When warm oil is standing in a tank, water that may be in it can evaporate and condense out upon the upper cooler surfaces of the tank not covered by oil. Rusting could take place and vibration may cause this rust to fall into the oil. Tanks should be given some protective type of coating to avoid rusting.

Drainings from scavenge spaces and stuffing boxes should not be put into the oil system and stuffing box and telescopic pipe glands must be maintained in good condition to prevent entry of water, fuel and air into the oil system.

Regular examination and testing of the main circulating oil is important. Samples should be taken from a pipeline in which the oil is flowing and not from some tank or container in which the oil is stationary and could possibly be stagnant.

Smelling the oil sample may give indication of fuel oil contamination or if acrid, heavy oxidation. Dark colour gives indication of oil deterioration, due mainly to oxidation. Dipping fingers into the oil and rubbing the tips together can detect reduction in oiliness – generally due to fuel contamination – and the presence of abrasive particles. The latter may occur if a filter has been incorrectly assembled, damaged or automatically by-passed. Water vapour can condense on the surfaces of sight glasses, thus giving indication of water contamination. But various tests are available to detect water in oil, $e.g.$ immersing a piece of glass in the oil, water finding paper or paste – copper sulphate crystals change colour from white to blue in the presence of water – plunging a piece of heated metal such as a soldering iron into the oil causes spluttering if water is present.

A check on the amount of sludge being removed from the oil in the purifier is important, an increase would give indication of oil deterioration. Lacquer formation on bearings and excessive carbon formation in oil cooled pistons are other indications of oil deterioration.

Oil samples for analysis ashore should be taken about every 1,000-2,000 hours (or more often if suspect) and it would be recommended that the oil be changed if one or more of the following limiting values are reached:

1. 5% change in the viscosity from new. Viscosity increases with oxidation and by contamination with heavy fuel, diesel oil can reduce viscosity.

2. 0·5% contamination of the oil.

3. 0·5% emulsification of the oil, this is also an indication of water content. Water is generally permissible up to 0·2%, dangerous if sea water.

4. 1·0% Conradson carbon value. This is from cracked lubricating oil or residue from incomplete combustion of fuel oil.

5. 0·01 mg KOH/g Total Acid Number (TAN). The TAN is the total inorganic and organic acid content of the oil. Sulphuric acid from engine cylinders and chlorides from sea water give the inorganic, oxidation produces the weak organic acids. Sometimes the acids may be referred to as Strong and Weak.

LUBRICATION SYSTEMS

Lubrication systems for bearing and guides, etc. should be simple and effective. If we consider the lubrication of a bottom end bearing, various routes are available, the object would be to choose that route which will be the most reliable, least expensive and least complicated. We could supply the oil to the main bearing and by means of holes drilled in the crankshaft convey the oil to the bottom end bearing. This method may be simple and satisfactory for a small engine but with a large diesel it presents machining and stress problems.

In one large type of diesel the journals and crankpins were drilled axially and radially, but to avoid drilling through the crank-web and the *shrinkage surfaces* the oil was conveyed from the journal to the crank pin by pipes.

A common arrangement, mainly adopted with engines having oil cooled pistons, is to supply the bottom end bearing with oil down a central hole in the connecting rod from the top end bearing. Fig. 39. shows an arrangement wherein a telescopic pipe-system is used and Fig.40. a swinging arm, the disadvantage of the latter is that it has three glands whereas the telescopic has only one. However, it is more direct and could be less expensive.

With any of the bearings (excepting ball or roller) the main object is to provide as far as possible a good hydrodynamic film of lubricant (*i.e.* a continuous unbroken film of oil separating the working surfaces). Those factors assisting hydrodynamic lubrication are:

1. *Viscosity*. If the oil viscosity is increased there is less likelihood of oil film break down. However, too high a viscosity increases viscous drag and power loss.

2. *Speed*. Increasing the relative speed between the lubricated surfaces pumps oil into the clearance space more rapidly and helps promote hydrodynamic lubrication.

3. *Pressure*. Increasing bearing load and hence pressure (load/area) breaks down the oil film. In design, if the load is increased area can be increased by making the pin diameter larger – this will also increase relative speed.

FIG 39
LUBRICATION OF BEARINGS

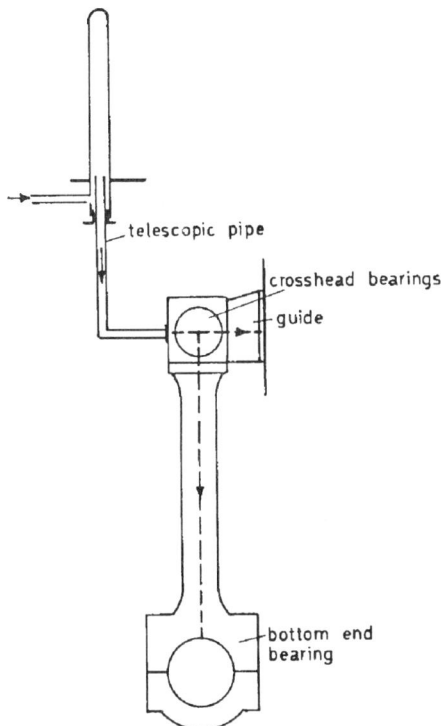

4. *Clearance.* If bearing clearance is too great inertia forces lead to 'bearing knock.' This impulsive loading results in pressure above normal and breakdown of the hydrodynamic layer. Fig. 41. illustrates the foregoing points graphically for a journal type of bearing.

Hydrodynamic lubrication should exist in main, bottom end and guide bearings. The top end bearing will have a variable condition, *e.g.* when at T.D.C. relative velocity between crosshead pin and bearing surface is zero and bearing pressure near or at maximum. Methods of improving top end bearing lubrication are:

1. Reversal of load on top end by inertia forces – only possible with medium or high speed diesels.

FIG 40
LUBRICATION SYSTEM FOR MAIN BEARINGS

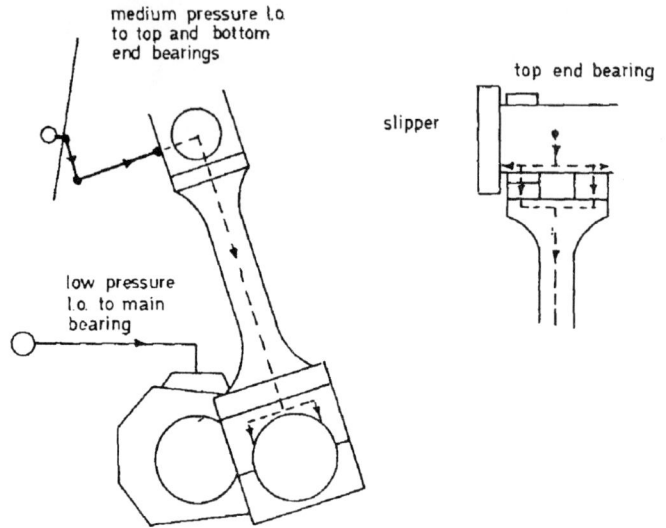

2. Use as large a surface area as possible, *i.e.* the complete underside of the crosshead pin. Fig. 42.

FIG 41
RELATIONSHIP BETWEEN COEFFICIENT OF FRICTION AND SURFACE SPEED

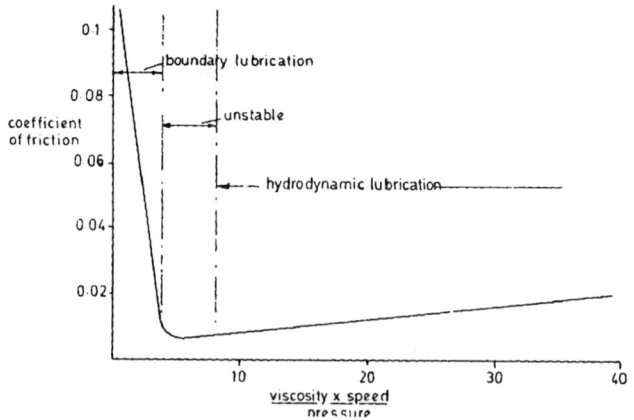

FIG 42
ONE PIECE LOWER BEARING CROSSHEAD DESIGN. SKETCH SHOWS LHS SLIPPER.

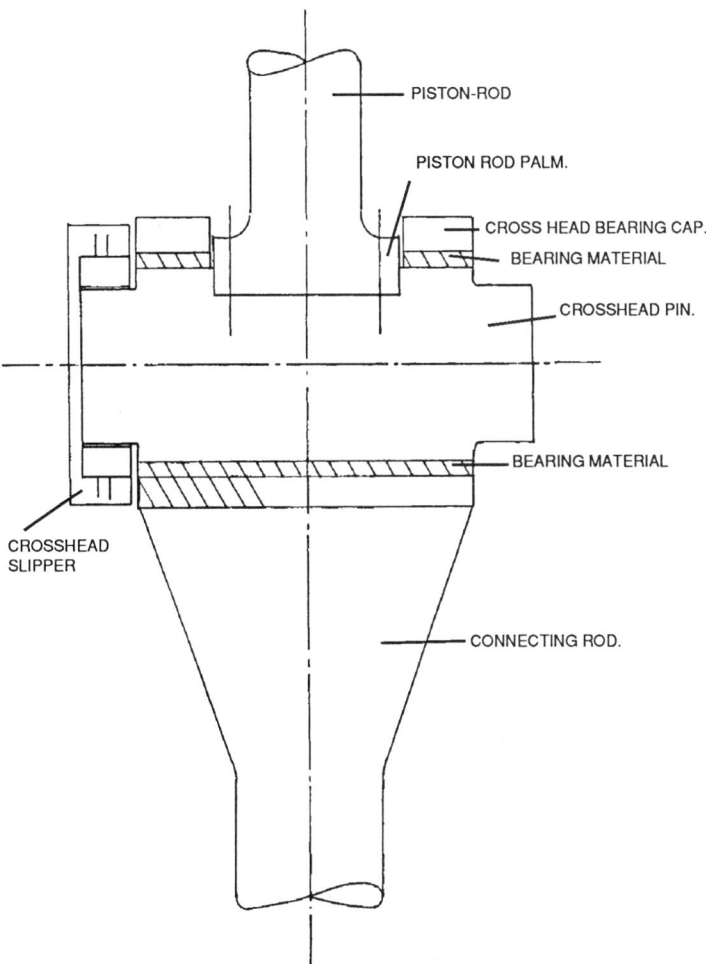

3. Avoid large axial variation of bearing pressure by more flexible seating and design. Fig. 43.

4. Increase oil supply pressure. Fig. 44. shows a method of increasing oil supply pressure to the top end bearing which tends to keep the crosshead pin 'floating' at all times. As the connecting rod oscillates, lubricating cross head oil is pumped at high pressure from the two pumps (only one is shown).

Increase in oil supply pressure can also be accomplished by installing lubricating oil booster pumps taking suction from L. O. system.

CYLINDERS AND PISTONS

Cylinders

Fig. 45. shows in section a cylinder liner from a large 2-stroke engine. The liner is manufactured from good quality lamellar cast iron and must satisfy the conflicting requirements of being thick and strong enough to withstand the high pressures and temperatures that occur during combustion and thin enough to allow good heat transfer.

This conflict is reconciled by the use of bore cooling. It can be seen in Fig. 46. that, by boring the upper part of the liner at an angle to the longitudinal axis the bore at mid point is close to the surface of the liner. The close proximity of the liner surface to the cooling water results in effective heat transfer. By using the technique of bore cooling good heat transfer is accompanied by high overall strength.

By maintaining the correct surface temperatures in the vicinity of the combustion space by good heat transfer, there is the risk of low temperature corrosion or cracking occurring in the lower portions of the liner. The solution to this problem is to either insulate the cooling water spaces that are at risk, or utilise a load controlled cylinder cooling system to maintain optimum cylinder liner temperature. Fig 131.

Longitudinal expansion of the liner takes place through the lower cooling jacket. The sealing of the cooling water is accomplished by silicone rubber "O" rings installed in grooves machined in the liner, which slide over the jacket as the liner expands and contracts. Fig. 45. The "O" rings are in groups of two, the space between them being open to the atmosphere. Leakage

FIG 43
CROSSHEAD WITH FLEXIBLE BEARING SUPPORTS

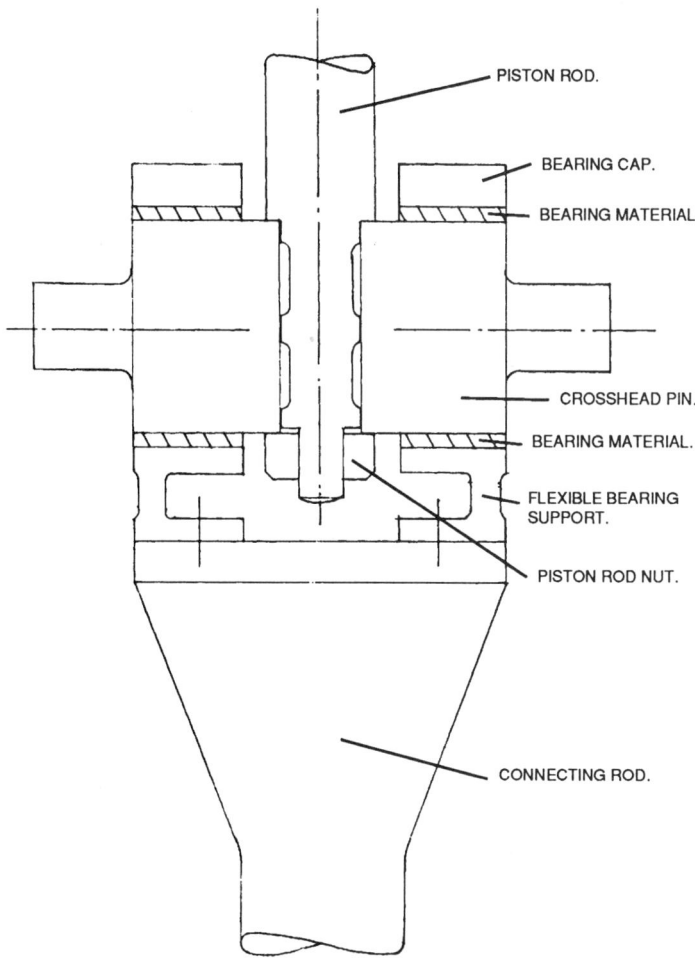

FIG 44
CROSS HEAD OIL PUMP

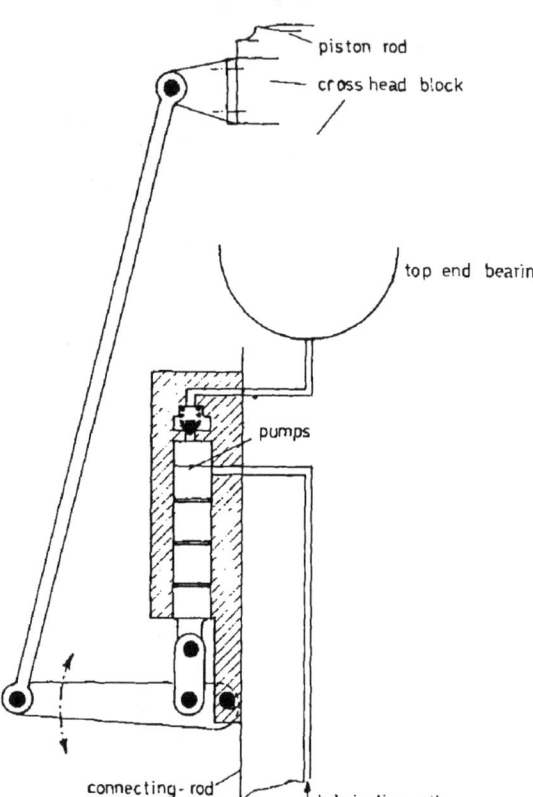

past an "O" ring will leak from this inspection hole not only alerting the engineers but preventing leakage into the scavenge space. Great care must be exercised to ensure that the "O" rings are not damaged when re-fitting a cylinder liner into the cylinder jacket.

Cylinder Lubrication
The principal objects of cylinder lubrication are:
    1. To separate sliding surfaces with an unbroken oil film.
    2. To form an effective seal between piston rings and cylinder liner surface to prevent blow past of gases.

STRUCTURE AND TRANSMISSION

FIG 45
2 STROKE CYLINDER LINER

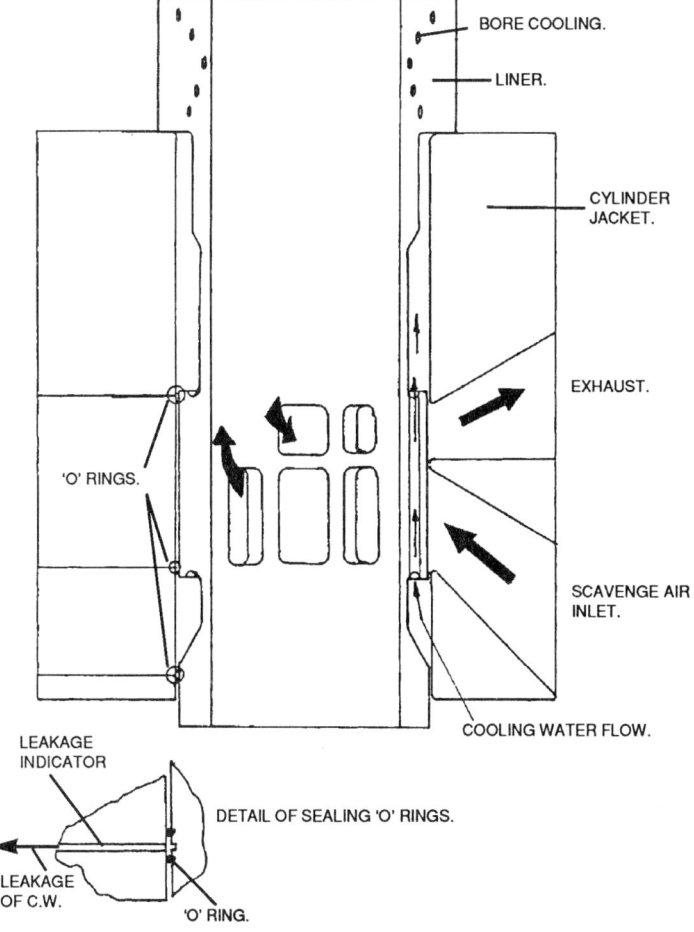

FIG 46
DIAGRAMMATIC VIEW OF CYLINDER LINER BORE COOLING

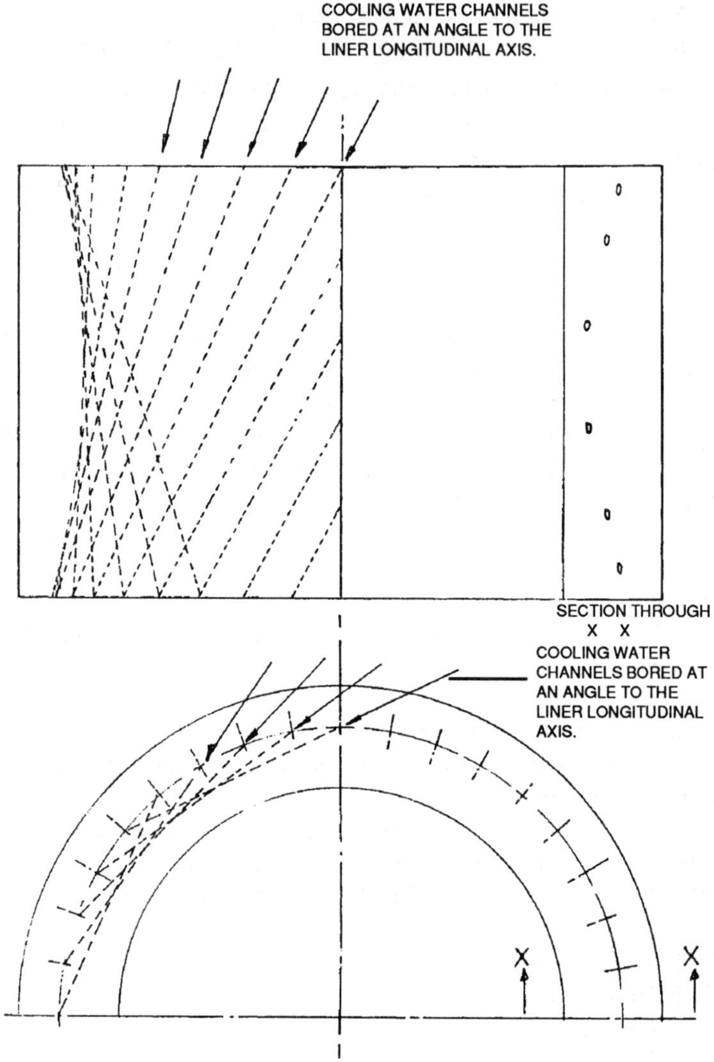

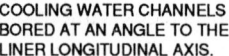

3. To neutralise corrosive combustion products and thus protect cylinder liner, piston and rings from corrosive attack.
4. To soften deposits and thus prevent wear due to abrasion.
5. To remove deposits to prevent seizure of piston rings and keep engine clean.
6. To cool hot surfaces without burning.

In practice some oil burning will take place, if excessive this would be indicated by blue smoke and increased oil consumption. As the oil burns it should leave as little and as soft a deposit as possible. Over lubrication should be avoided.

When the engine is new, cylinder lubrication rate should normally be greater than when the engine is run in. Reasons for this increased lubrication are: (1) surface asperities will, due to high local temperatures, cause increased oxidation of the oil and reduce its lubrication properties, (2) sealing of the rough surfaces is more difficult, (3) worn off metal needs to be washed away.

The actual amount of lubricating oil to be delivered into a cylinder per unit time depends upon: stroke, bore and speed of engine, engine load, cylinder temperature, type of engine, position of cylinder lubricators and type of fuel being burnt.

Position of the cylinder lubricators for injection of oil has always been a topic of discussion, the following points are of importance:

1. They must not be situated too near the ports, oil can be scraped over edge of ports and blown away.
2. They should not be situated too near the high temperature zone or the oil will burn easily.
3. There must be sufficient points to ensure as even and as complete a coverage as possible.

Ideally, timed injection of lubricant delivering the correct measured quantity to a specific surface area at the correct time in the cycle is the aim, but is difficult to achieve in practice.

Cylinder Liner Wear

Cylinder liner wear can be divided into:
- Abrasive wear.
- Corrosive wear.

Abrasive wear
This occurs when abrasive particles enter the combustion space with scavenge air or as a result of poor quality or contaminated fuel. Instances of extremely high abrasive wear rates have occurred in the past due to the burning of fuel heavily contaminated by catalytic fines.

Corrosive wear
This is the more common cause of cylinder liner wear, caused when burning heavy fuel which contains significant amounts of sulphur. As the fuel burns the sulphur combines with oxygen to produce oxides of sulphur which form sulfuric acid on contact with water. To minimise the formation of acids it is important that cylinder liner temperatures are maintained above the dew-point. Fig. 47.

To minimise cylinder liner wear it is imperative that ship's engineers operate the engine correctly. This includes:
- Correct quantity and grade of cylinder lubrication.
- Correctly fitted piston rings.
- Correct warming through prior to starting.
- Well maintained and timed fuel injectors.
- Well managed fuel storage and purification plant.
- Correct cooling water and lubricating oil temperatures.
- Correct scavenge air temperatures.
- Engine load changes carried out gradually.
- Well maintained equipment.

The deterioration of fuel quality that has taken place years coupled with the increased pressures and temperatures that occur during the combustion process have resulted in liners and piston rings operating under very severe conditions. Despite these adverse operating conditions cylinder liner wear rates have been reduced in recent years with large 2-stroke manufacturers claiming 0·03 mm/1000 hrs and medium speed 4-stroke engine manufacturers claiming wear rates of 0·02 mm/1000 hrs when operating on heavy fuel.

These wear rates have been achieved as a result of a number of factors such as:
- The development of highly alkaline lubricating oils to neutralise the acids formed during combustion.

FIG 47
TEMPERATURE OF CYLINDER LINER SURFACE THROUGHOUT.
ENGINE LOAD RANGE.

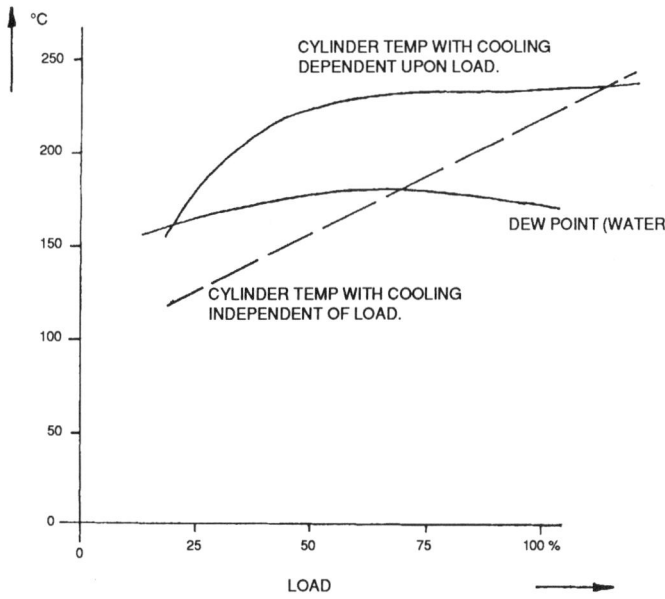

- The development of load dependent temperature control of cooling water which maintains the cylinder liner temperature at optimum level. Fig. 131 [cooling water section].
- The use of good quality cast iron with sufficient hard phase content for cylinder liners.
- Careful design of piston ring profiles to maximise lubricating oil film thickness.
- Improvements in lubricating oil distribution across cylinder liner surface. This includes multi-level injection in 2-strokes engine and forced piston skirt lubrication in 4-stroke engines. Fig. 51.
- Improved separation of condensate from scavenge air.

Cylinder liner wear profile
Fig. 48. shows the wear profile of both a 4-stroke and 2-stroke engine cylinder liner. It can be seen that the greatest wear occurs in the upper part of the liner adjacent to firing zone. This is due to:
• The high temperatures and pressures that occur at this point.
• Because the piston reverses direction at this point hydrodynamic lubrication is not established.
• Acids formed during combustion attack the liner material.

Cloverleafing
Despite the close control of cylinder surface temperatures, acids are still formed which must be neutralised by the cylinder lubricating oil. This requires that the correct quantity and TBN grade of oil is injected into the cylinder. Immediately the oil enters the cylinder it will start neutralising the acids, becoming less alkaline as it does so. If the TBN of the oil is too low then its alkalinity may be depleted before it has completely covered the liner surface. Further contact with the acids may lead to the oil itself becoming acidic. This will lead to the phenomenon known as "cloverleafing" in which high corrosive wear occurs on the liner between the oil injection points Fig. 49. Severe cloverleafing can result in gas blow-by past the piston rings and ultimate failure of the liner.

Micro-seizure
This is due to irregularities in the liner and piston rings coming into contact during operation as a result of a breakdown of lubrication due to an insufficient quantity of lubricating oil, insufficient viscosity or excessive loading. This results instantaneous seizure and tearing taking place. In appearance micro-seizure resembles abrasive wear since the characteristic marks run axially on the liner. Micro-seizure may not always be destructive, indeed it often occurs during a running-in period. It becomes destructive if is persistent and as a result of inadequate lubriation.

PISTONS AND RINGS
Pistons
Pistons must be strong enough to withstand the very high firing pressures that are common today, be able to dissipate sufficient

FIG 48
LINER WEAR PROFILE

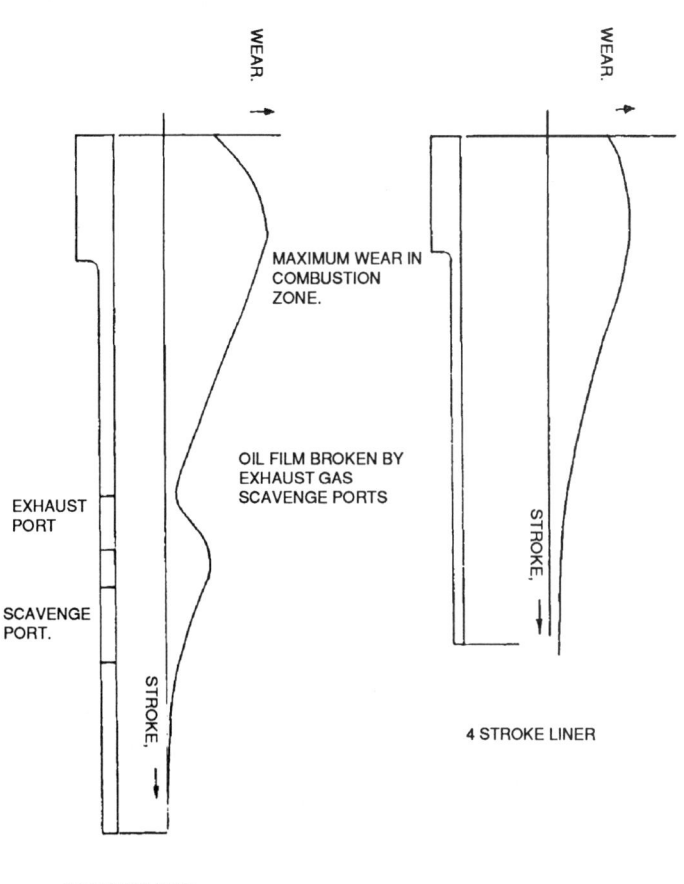

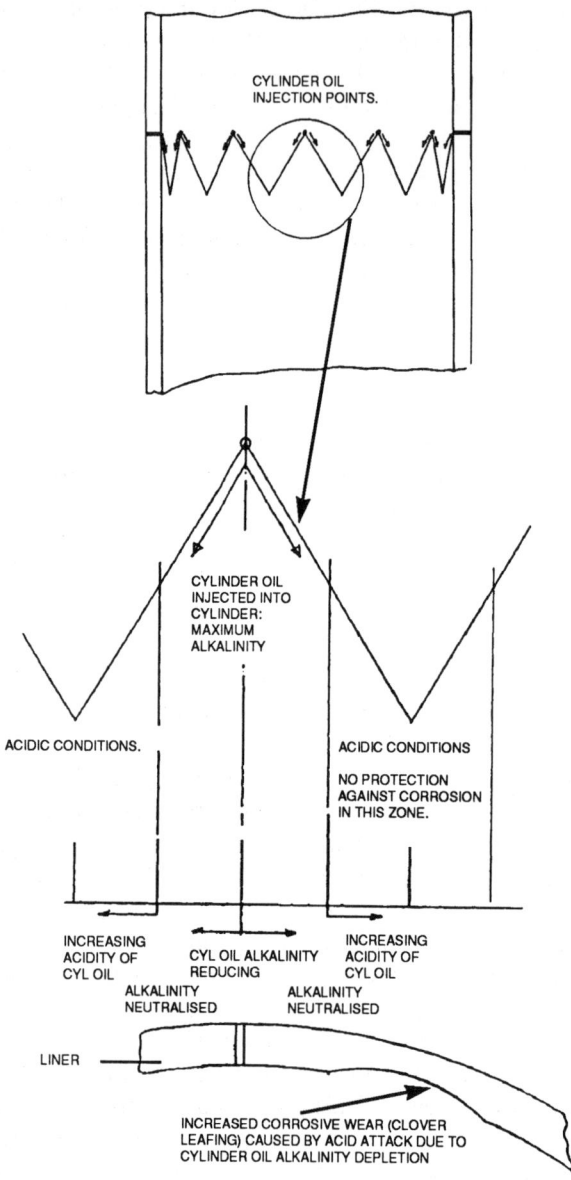

**FIG 49
INCREASED CORROSIVE WEAR OF CYLINDER LINER. (CLOVERLEAFING).**

heat to maintain the correct piston crown temperatures and withstand the stresses imposed by friction. Pistons are manufactured from cast steel, forged steel, and cast iron although all of these materials have limitations. Cast iron is weak in tension especially at elevated temperatures. It does, however, have high compressive strength which enables it to resist the hammering which occurs at the ring grooves. Because of its graphite content, cast iron performs well when exposed to rubbing. This makes it a suitable material for piston skirts. Cast steel resists heat stresses better than cast iron but is difficult to ensure that the molten material flows to the extremities of intricate moulds. Cast steel also requires extensive heat treatment to relieve casting stresses. Forged steel is a suitable material because the directional grain flow exhibited as a result of forging produces a strong tough component. Forged steel is prone to high wear at the ring grooves and also requires a greater degree of machining which tends to increase the production cost.

Modern pistons are composite components, made from materials that exhibit suitable properties.
- Piston crowns which are highly stressed mechanically and thermally are made from cast or forged steel. Cast iron inserts are fitted to the ring grooves to resist wear.
- Piston skirts are made from cast iron which has superior rubbing properties than either cast or forged steel. To reduce weight and reduce inertia loads aluminium is used in some medium speed 4-stroke applications.

Fig. 50. shows a piston for a large Sulzer engine. The crown is of forged steel and combines strength with good heat transfer.

Strength is achieved by using an overall thick section piston crown which is then bore cooled. Intensive cooling is achieved by the cocktail shaker effect of the water. With air present in the piston [this comes from the telescopic system, it being necessary to provide a cushion and prevent water hammer] together with water, the inertia effect coupled with the bore cooling leads to very effective cooling as the piston goes over TDC.

Fig. 51. shows an oil cooled piston of a large modern B.& W. engine. The piston crown of this piston is also manufactured from forged steel but in this case the section is relatively fine. Strength being achieved by the "strong back" principle which supports the

FIG 50
WATER COOLED PISTON WITH BORE COOLING

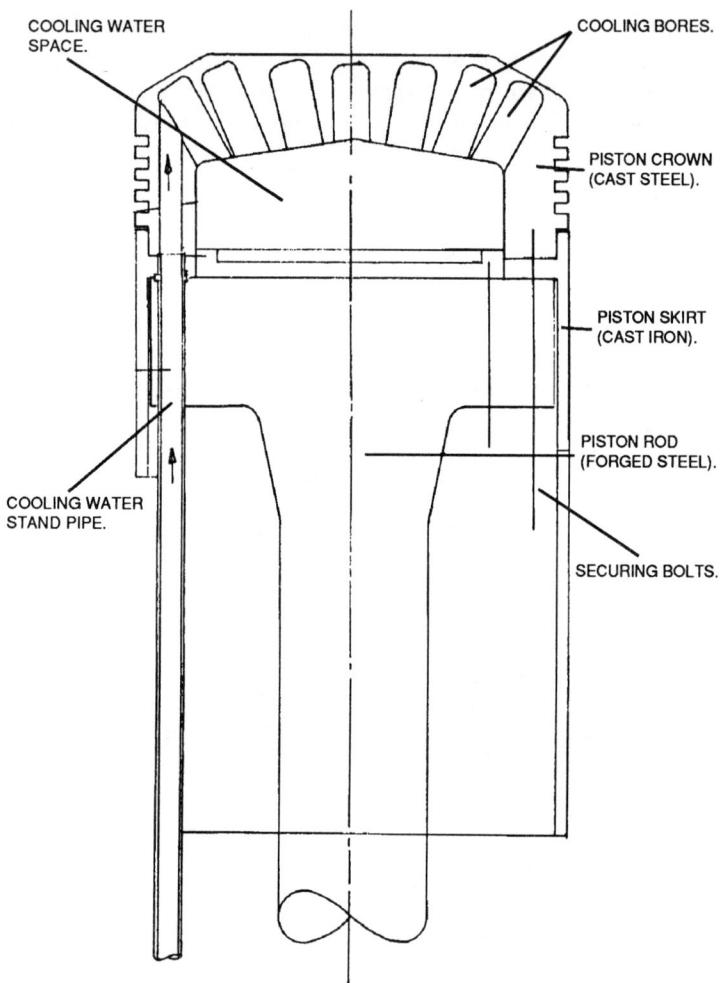

FIG 51
OIL COOLED PISTON FROM 2 STROKE ENGINE.

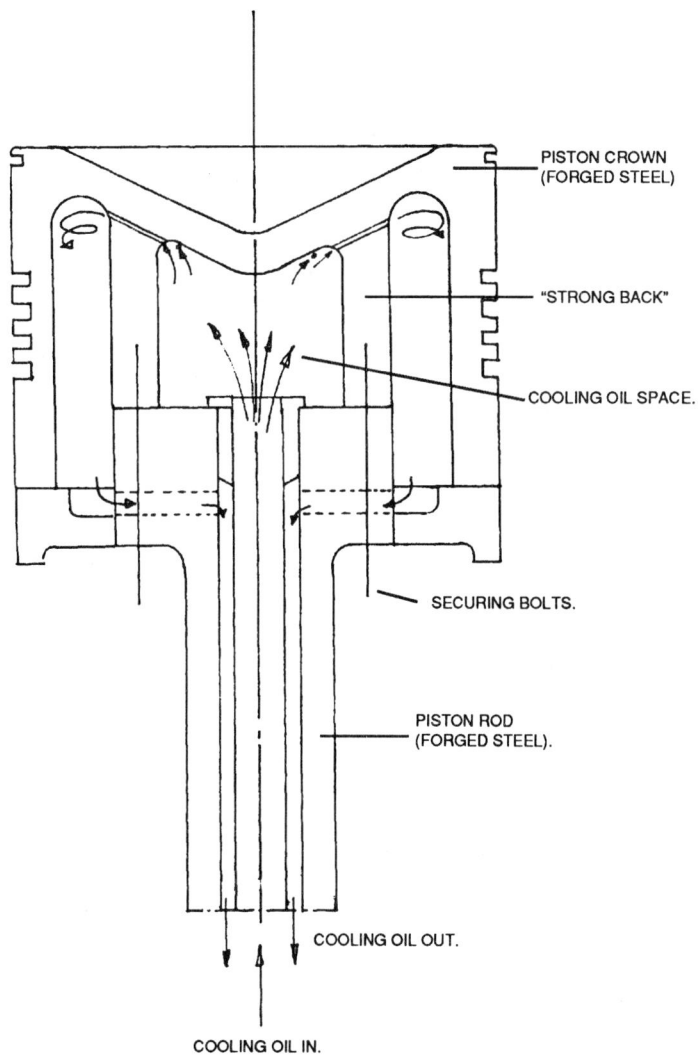

piston crown from inside. Bore cooling is employed but in this design the bores act as nozzles through which the oil flows radially, spraying onto the underside of the piston crown, before flowing to the drain.

Fig. 52. shows a piston from a medium speed 4-stroke engine. This design transmits the combustion forces directly onto the gudgeon pin. The piston crown is of forged or cast steel whilst the skirt is of nodular cast iron. Cooling is effected by oil flowing from the connecting rod into the piston crown then flowing radially outwards to effectively cool the piston. In Wartsilla engines some of this oil is then taken out through four nozzles which feed the oil distribution groove in the piston skirt. The manufacturers claim that this design, which they have patented, provides an even oil film formation that reduces liner wear.

The Choice of water or oil for piston cooling.
Distilled water, kept free from impurities and in the correct alkaline state, has some advantages over oil:
- It is relatively cheap and plentiful.
- Internal surfaces are kept free from deposits.
- Water can be operated at higher temperatures.
- Water has a specific heat capacity nearly twice that of oil.
  [This means that, for the same mass flow rate, water is able to transfer nearly twice as much heat away as oil, lower mass flow rates can be specified.]

The disadvantages of water are that:
- Leakage into the crankcase will result in serious contamination of the lubricating oil.
- Additional pumps and coolers are required.

Oil is used extensively in modern engines. The advantages claimed for this medium are:
- Simplified supply of the oil to the piston is achieved.
- Leakage into the crankcase does not present contamination problems.
- The lower thermal conductivity results in a less steep temperature gradient over the piston crown.

FIG 52
COMPOSITE PISTON SUITABLE FOR A HIGH OUTPUT MEDIUM SPEED 4-STROKE DIESEL ENGINE.

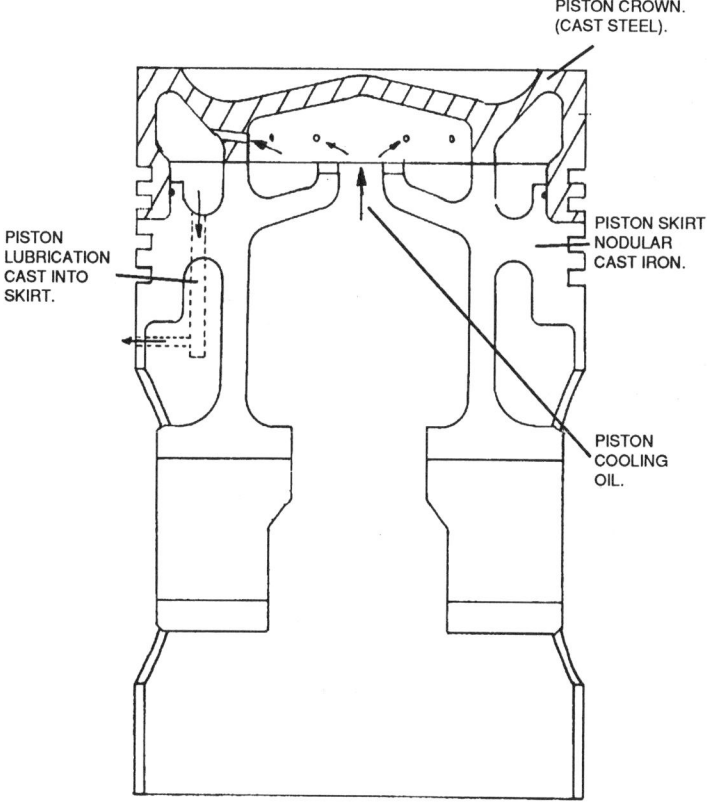

The disadvantages of oil are:
- The temperatures must be kept relatively low in order to limit oxidation of the oil.
- If overheating occurs there is a a possibility that carbon deposits could form on internal surfaces and the danger that carbon particles could enter the lubricating oil system.

Failure of pistons due to Thermal Loads

When a piston crown is subjected to high thermal load, the material at the gas side attempts to expand but is partly prevented from doing so by the cooler metal under and around it. This leads to compressive stresses in addition to the stresses imposed mechanically due to the variation in cylinder pressures.

At very high temperatures the metal can creep to relieve this compressive stress and when the piston cools a residual tensile stress is set up hence residual thermal stress. If this stress is sufficiently great, cracking of the piston crown may result.

At normal working temperatures the piston and cylinder liner surfaces should be parallel. Since there is a temperature gradient from the top to the bottom of the piston, allowance must be made during manufacture for the top cold clearance to be less than the bottom. The temperature gradient is generally non-linear and thermal distortions produce tensile stresses on the inner wall of the piston, gas forces tend to bulge the piston wall out thereby reducing the tensile stress. This variable tensile stress at very high thermal loads could lead to cracks propagating through from the inside of the piston to the piston ring grooves.

FIG. 53
EFFECT OF GAS AND HEAT

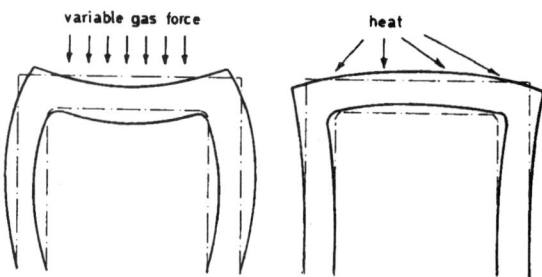

Piston Rings

Properties required of a piston ring:
 1. Good mechanical strength, it must not break easily.
 2. High resistance to wear and corrosion.
 3. Self lubricating.
 4. Great resistance to high temperatures.
 5. Must at all times retain its tension to give a good gas seal.
 6. Be compatible with cylinder liner material.

The above properties are the ideal and therefore difficult to achieve in practice. Materials that are used to obtain as many of the desired properties as possible are as follows:

 1. Ordinary grey cast iron, in order that it may have good wear resistance and self lubricating property it must have a large amount of graphite in its structure. This however reduces its strength.

 2. Alloyed cast iron, elements and combinations of elements that are alloyed with the iron to give finer grained structure and good graphite formation are: Molybdenum, Nickel and Copper or Vanadium and Copper.

 3. Spheroidal Graphitic iron, very good wear resistance, not as self lubricating as the ordinary grey cast iron. These rings are usually given a protective coating, *e.g.* chromed or aluminised, etc. to improve running-in.

It is possible to improve the properties by treatment. In the case of the cast irons with suitable composition they can be heat treated by quenching, tempering or austempering. This gives strength and hardness without affecting the graphite.

Piston rings are often contoured to assist in the establishment of a hydrodynamic lubricating oil film and so reduce liner wear Fig. 54. It is common practice for manufacturers to specify a ring pack in which the first and second compression rings, subjected to higher temperatures and pressures, differ from the lower rings. It is important when installing new rings that the manufacturers recommendations are followed since ring failure may result if incorrect rings are fitted.

In addition to compression rings 4-stroke medium speed engines also employ oil control, or oil scraper rings Fig. 55. Unlike compression rings, which help promote the formation of an oil film, oil scraper rings scrape the oil from the cylinder liner and return it to the sump. Many designs of oil scraper rings can only be

FIG 54
2 STROKE ENGINE PISTON RING PROFILE.

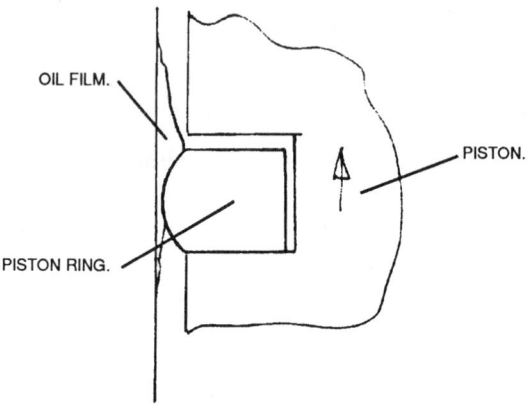

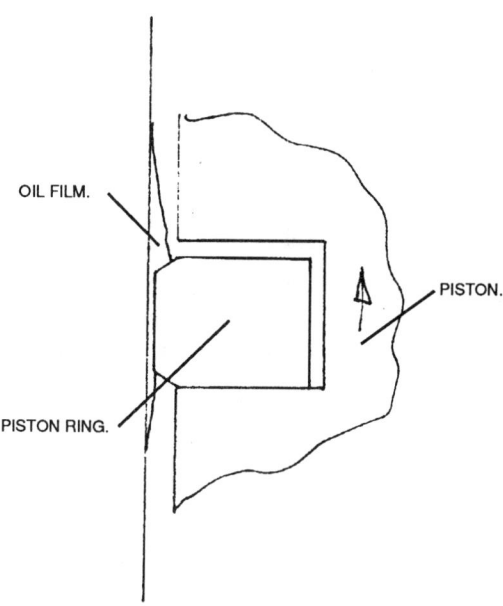

FIG 55
4 STROKE PISTON RINGS.

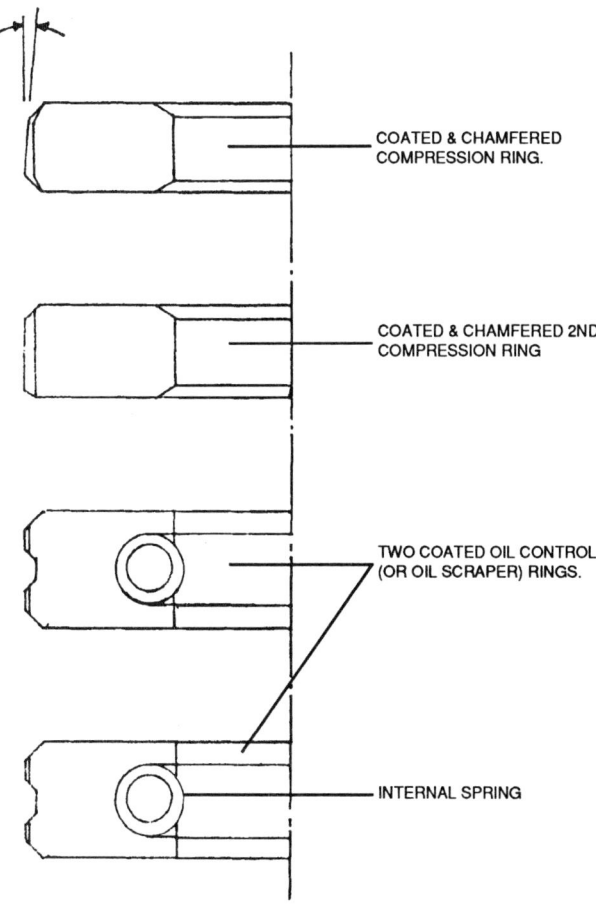

fitted in one direction and care must be exercised when installing these rings. Without these rings lubricating oil in the upper cylinder would be burnt during combustion resulting in extremely high oil consumption. As the oil scraper rings wear their effectiveness in returning the oil to the sump reduces with high oil consumption as the consequence. Oil scraper ring wear may be the limiting factor when deciding cylinder overhaul periods for medium speed 4-stroke engines.

Manufacture

1. Statically cast in sand moulds to produce either a drum from which a number of piston rings would be manufactured or an individual ring.

2. Centrifugally cast to produce a fine grained non-porous drum of cast iron from which a number of piston rings will be machined. The statically cast rings, either drum or single casting, may be made out of round. The out of round blanks are machined in a special lathe that maintains the out of roundness. Rings manufactured in this way are expensive but ideal.

Most piston rings are made from circular cast blanks which are machined to a circular section on their inner and outer diameters. In order that the rings may exert radial pressure when fitted into the cylinders they are split in tension. Tensioning is done by cold deformation of the inner surface by hammering or rolling. The finished ring would be capable of exerting a radial pressure from 2 to 3 bar and have a Brinel hardness from 1600 to 2300 (S.I. units). Large diesel engine cylinder liners have a hardness range similar to the above.

Piston ring defects and their causes

1. Incorrectly fitted rings. If they are too tight in the grooves the rings could seize causing overheating, excessive wear, increased blow-past, etc. If they are too slack in the grooves angular working about a circumferential axis could cause ring breakage and piston groove damage. If the butt clearance is too great, excessive blow-past will occur.

2. Fouling due to deposits on the ring sides and their inner diameters, this could lead to rings sticking, breakage, increased blowpast and scuffing.

3. Corrosion of the piston rings can occur due to attack from corrosive elements in the fuel ash deposits.

4. If the ring bearing surfaces are in poor condition or in any way damaged (this could occur during installation) scoring of cylinder liner may take place, if the ring has sharp edges it will inhibit the formation of a good oil film between the surfaces.

Due to uneven cylinder liner wear the piston ring diameter changes during each stroke, this leads to ring and groove wear on the horizontal surfaces. This effect obviously increases as differential cylinder liner wear increases. Oscillation of the piston rings takes place in the cycle about a circumferential axis approximately through the centre of the ring section, and if the inner edges are not chamfered they can dig into the piston groove lands. Keeping the vertical clearance to a working minimum will reduce the oscillatory effect.

When considering piston rings perhaps the most destructive force at work is hammering. This is caused by relative axial movement between piston and ring as a result of gas loading and inertia when the piston changes direction at BDC. The hammering results in enlargement of the piston ring groove and may result in ring breakage. Cast steel, forged steel or aluminium pistons usually have ring groove landing surfaces protected to minimise the effects of hammering. This can be either:
- Flame hardening - top and bottom on upper grooves.
- Chromium plating- top and bottom on upper grooves.
- The fitting of cast iron inserts.

In 2-stroke engines, piston rings have to pass ports in the cylinder wall. Each time they do, movement of segments of the rings into the ports can take place. This would be more pronounced if the piston ring butts are passing the ports. It is possible for the butts to catch the port edge and bend the ring. In order to avoid or minimise this possibility, piston rings may be pegged to prevent their rotation or they may be specially shaped.

Inspection of pistons, rings and cylinders
Withdrawal of pistons their examination, overhaul or renewal, together with the cleaning and gauging of the cylinder liner, is a regular feature of maintenance procedures. Frequency of which depends upon numerous factors, such as: piston size, material and

method of cooling; engine speed of rotation; type of engine, 2- or 4-stroke; fuel and type of cylinder lubricant used.

With high speed diesel engines of the 4-stroke type running time between piston overhauls is generally greater than that for large slow running two stroke engines. This can be attributed to the facts that: the engine is usually unidirectional, hence reduced numbers of stops and starts with their attendant wear and large fluctuations of thermal conditions. Small bore engines are easier to cool, cylinder volume is proportional to the square of the cylinder diameter hence increasing the diameter gives greatly increased cylinder content and high thermal capacity. Thus overhaul time can vary between about 2000 to 20 000 h.

Pistons and cylinder liners on some engines can be inspected without having to remove the piston.

After scavenge spaces have been cleaned of inflammable oil sludge and carbon deposits, each piston can in turn be placed at its lowest position. The cylinder liner surfaces can then be examined with the aid of a light introduced into the cylinder through the scavenge ports. The cylinder liner surfaces should have a mirror-like finish. However, black dry areas at the top of the liner indicate blow past of combustion gases. Dull vertically striped areas indicate breakdown of oil film and hardened metal surface (this is caused by metal seizure on a micro scale leading to intense heating).

After inspection of a cylinder the piston can be raised in steps in order to examine both the piston and the rings. Heavy carbon deposits on piston crown and burning away of metal would indicate incorrect fuel burning and poor cooling. Piston rings should be free in the grooves, have a well oiled appearance, be unbroken and worn smooth and bright on the outer surface. If they are too worn then sharp burrs can form on the edges which enable them to act as scraper rings, preventing good oil film formation.

Large 2-stroke engines
The cylinder covers of loop scavenged engines tend to be relatively simple symmetrical designs to avoid the problems of differential expansion and the consequent stresses. Early Sulzer designs were two piece with a cast iron main component and a central cast steel insert containing the valves. In this design cooling water is introduced into the cylinder cover through nozzles which ensure

that the water flows tangentially thus minimising impingement on internal surfaces thus reducing the possibility of erosion Fig. 56. More modern engines, which operate at higher temperatures and pressures, have one piece forged steel cylinder covers. This design employs bore cooling which allows the cooling water to pass very close to the combustion chamber effectively maintaining safe surface temperatures. Fig. 57.

FIG. 56
TWO PIECE CYLINDER COVER:
SULZER RND TYPE.

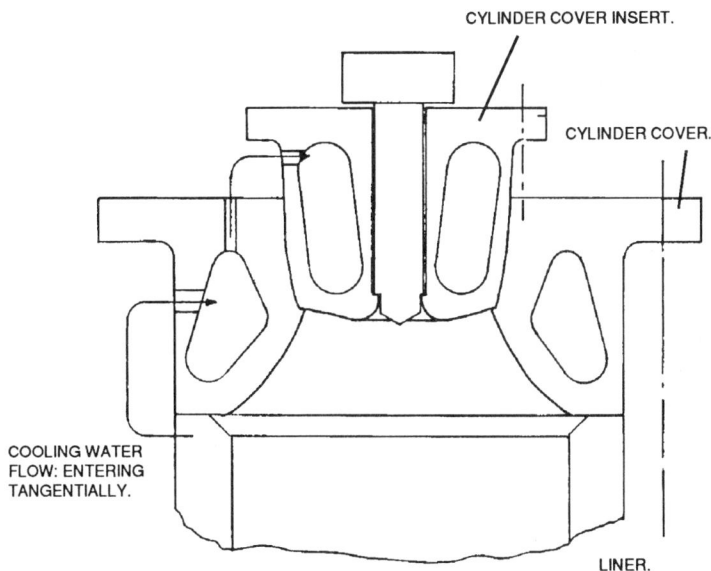

**FIG. 57
ONE PIECE CYLINDER COVER
SULZER RND M.**

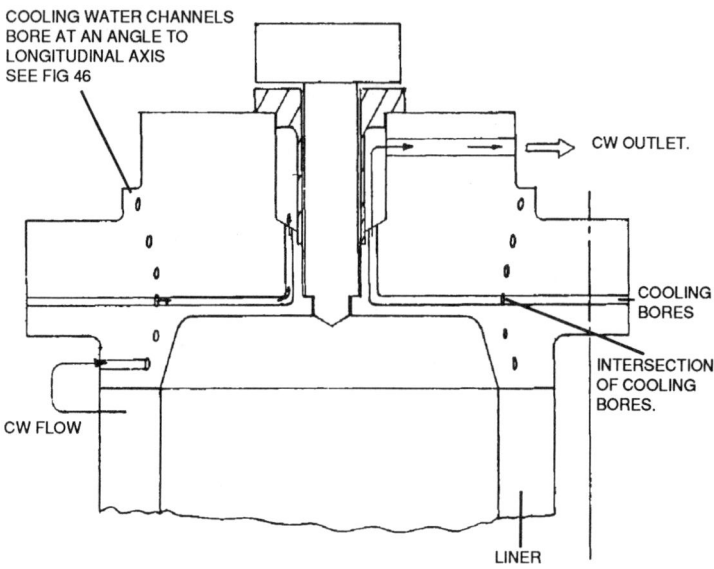

With the exception of opposed piston configurations uniflow engines have a single central exhaust valve in removable casting installed in the cylinder cover. Fig. 58. The cylinder cover is manufactured from forged steel and cooling is accomplished through radial cooling bores close to the combustion chamber surface. The exhaust valve cage and seat ring are also bore cooled.

FIG. 58
HYDRAULICALLY ACTIVATED CENTRAL EXHAUST VALVE FOR LARGE SLOW-SPEED 2-STROKE DIESEL ENGINE

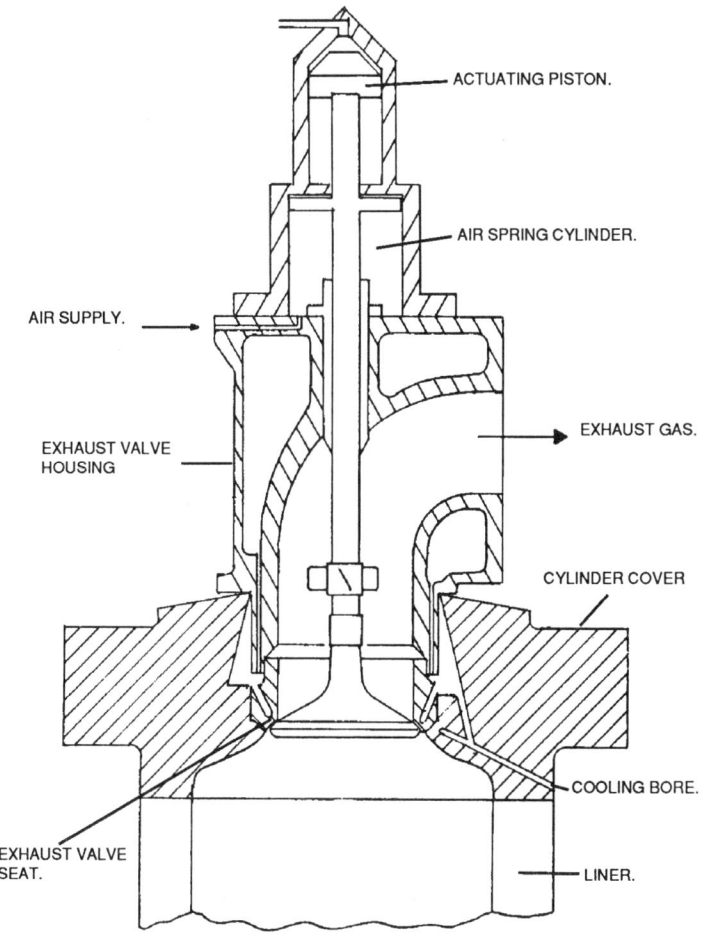

Exhaust valves

In recent years manufacturers have adopted the hydraulically actuated valves in favour of mechanical pushrods and rockers. The advantages claimed for this configuration are:
- There is no transverse thrust from hydraulic actuators. Thrust is purely axial resulting in less guide wear.
- Controlled landing speed ensures minimum stress on valve and seat.
- Valve rotation by impeller ensures well balanced thermal and mechanical stress and uniform valve seating.

The extensive cooling of the valve cage and seating ring results in relatively low exhaust valve seat temperatures which coupled with the choice of Nimonic for the one piece valve increases reliability and the intervals between overhauls even when operating on heavy fuel.

Medium speed 4-stroke engines

Because of the number of openings required for valves, 4-stroke cylinder heads are a complex shape. For this reason spheroidal graphite cast iron is a suitable material since it is relatively easy to cast. A modern cylinder head for an engine operating on heavy fuel is shown in Fig. 59. Such a cylinder head should:
- Have even and small thermal and mechanical deformation with correspondingly low stress levels.
- Have low and uniform temperature distribution at the exhaust valves and valve seat.
- Have good valve seating due to effective valve rotation and low levels of distortion [important for optimum heat transfer from exhaust valve seats].
- Have exhaust valves made from a material that provides resistance to high temperature corrosion.

Effective cooling of the exhaust valve and seat must be accomplished if reliable operation is to be achieved.

To maintain low surface temperatures in the combustion space and at the valve seat bore cooling is employed. The bore cooling passages are shown in Fig. 59.

Four valves are usually employed on 4-stroke engines. This configuration allows the designer to maximise the cross-sectional area of inlet and exhaust ports and so improve the flow through the cylinder. This arrangement results in more complicated valve

FIG 59
4 STROKE DIESEL ENGINE CYLINDER COVER WITH BORE COOLING

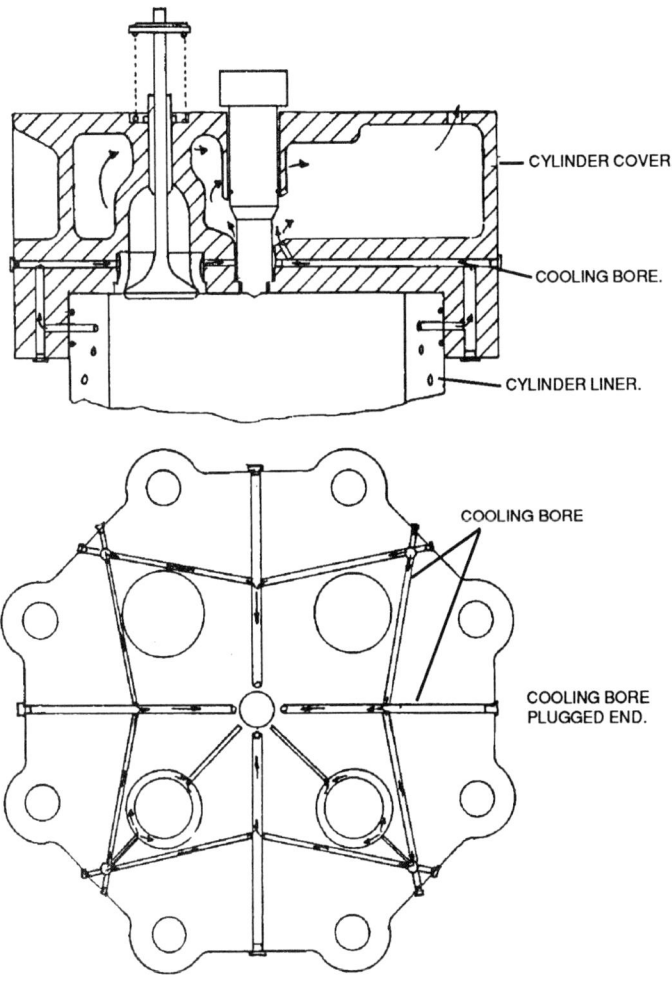

actuation since two exhaust and two inlet valves must each be operated by one push rod. The various ways that the valves are actuated can be seen in Fig. 60. All of the designs shown control the valves together, it is important that, following maintenance, adjustments are made correctly. Clearance is allowed between the valve stem and the rocker arm when the engine is cold. As the engine attains normal running temperature this clearance is taken up by expansion. If adjustments leave too little clearance then it is likely that the valve will be prevented from closing correctly by the valve gear, resulting in gas leakage, burning and deteriorating performance. Conversely, too great a clearance may result in reduced valve lift and duration of opening, mechanical noise and reduced performance levels.

STRUCTURE AND TRANSMISSION 99

**FIG 60
ALTERNATIVE METHODS OF VALVE
ACTUATION**

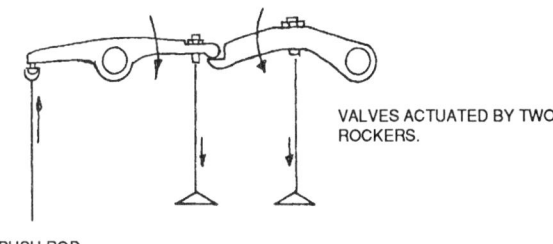

VALVES ACTUATED BY TWO ROCKERS.

PUSH ROD.

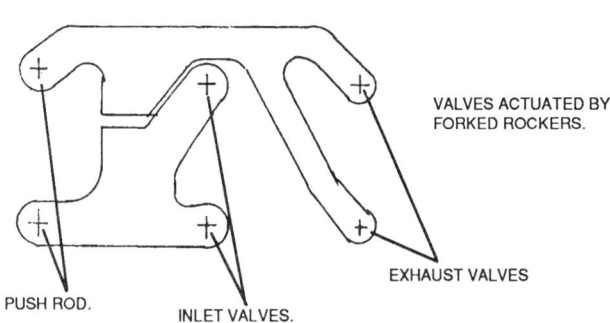

VALVES ACTUATED BY FORKED ROCKERS.

PUSH ROD.   INLET VALVES.   EXHAUST VALVES

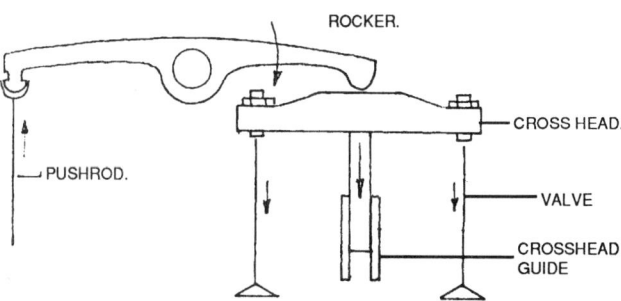

ROCKER.
CROSS HEAD.
PUSHROD.
VALVE
CROSSHEAD GUIDE

CHAPTER 3

FUEL INJECTION

DEFINITIONS AND PRINCIPLES

Atomisation
The break up of fuel into minute spray particles so as to ensure an intimate mixing of air and fuel oil is known as atomisation. The surface area/volume ratio of a fuel-oil droplet increases as its diameter decreases Fig. 61. The effect of this is that a smaller droplet can present a greater percentage of its molecules to contact with the available air than can a larger droplet. The smaller fuel-oil droplets then the more effective is the atomisation resulting in more rapid and complete combustion with maximum heat release from the fuel.

Turbulence
A swirl effect of air charge in the cylinder which in combination with atomised fuel spray gives intimate mixing and good overall combustion. Requires to be 'designed into' the engine by attention to liner, piston, ports, etc., details together with air pressure-temperature gradients.

Penetration
Ability of the fuel spray droplets to spread across the cylinder combustion space so as to allow maximum utilisation of volume for combustion.

Impingement
Excess velocity of fuel spray causing contact with metallic engine parts and resulting in flame burning.

Sprayer Nozzle
The arrangement at the fuel valve tip to direct fuel in the proper direction with the correct velocity. If the sprayer holes are too short the direction can be indefinite and if too long impingement can

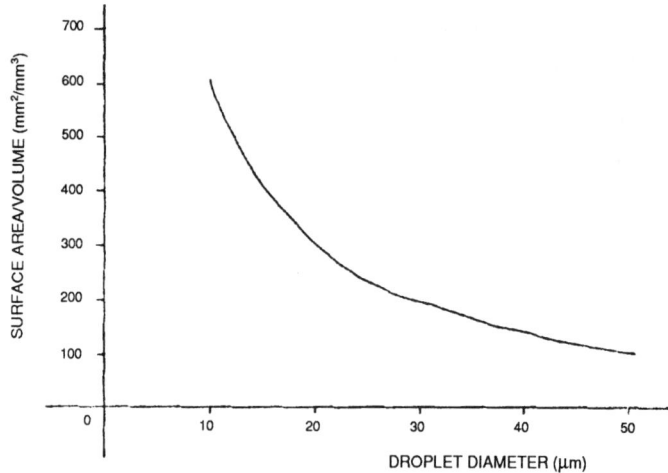

**FIG 61
THE RELATIONSHIP BETWEEN FUEL DROPLET SIZE & SURFACE AREA/VOLUME RATIO**

occur. If the hole diameters are too small fuel blockage (and impingement) can take place, alternatively too large diameters would not allow proper atomisation. In practice each manufacturer has a specific design taking into account method of injection, pressure, pumps, etc. Even with a particular engine different nozzles may be specified for different applications. For example, engines engaged in slow steaming, for reasons of economy, may be supplied with fuel valve sprayer nozzles with smaller holes of differing geometry than engines at higher powers. This measure improves the atomising and penetration performance of fuel valves at part load due to the restoration of fuel velocity through the nozzle. This yields improved economy at lower loads. The nozzles must be changed to the original size prior to operating at maximum power. As a generalisation the sprayer hole length: diameter ratio will be about 4:1, maximum pressure drop ratio about 12:1 and fuel velocity through the hole about 250m/s.

Viscosity
May be defined as internal fluid molecular friction which causes a resistance to flow.

Pre-heating
With the almost universal use of residual fuel in marine diesel engines it is important, to ensure optimum fuel injection, that correct pre-heating is carried out. If the temperature of the fuel is too low then the viscosity will be high resulting in higher injection pressure and reduced atomising performance, excessive penetration and possible impingement on internal surfaces. Too high a temperature also has adverse effects by reducing penetration and causing deposits to be left on nozzle tip affecting atomisation. The relationship between viscosity and temperature is shown in Fig. 62. Careful control of fuel temperature is required to ensure that the fuel viscosity at the engine fuel rail is inside the range specified by the manufacturers. It is modern practice to utilise viscosity controllers to ensure that the fuel temperatures is maintained at the correct level. Although the viscosity controller manufacturer will specify that it be fitted in close proximity to the engine the builders do not always do so. It is important, therefore, that the viscosity controller is adjusted so that any cooling of the fuel that takes place between the heater and the engine does not allow the fuel to move out of the optimum viscosity range for injection. This effect will be exacerbated when burning high viscosity fuels requiring much heating. Effective trace heating and insulation is an important feature of a well designed fuel system.

The fuel injection system is vitally important to the efficient operation of the engine. It must:
- Supply an accurately measured amount of fuel to each cylinder regardless of load.
- Supply the fuel at the correct time at all loads with rapid opening and closing of the fuel valve.
- Inject the fuel at a controlled rate.
- Atomise and distribute the fuel in the cylinder.

Injection
To achieve effective combustion the fuel must be atomised and then distributed throughout the combustion space. It is the function of the fuel valve to accomplish this. Most fuel valves on marine diesel engines are of the hydraulic type. Fig. 63. The opening and closing of this type of valve is controlled by the fuel pressure delivered by the fuel pump which acts on the needle in the lower chamber. When the force is sufficient to overcome the spring, the needle lifts.

FIG 62
VISCOSITY TEMPERATURE CHART FOR MARINE FUELS.

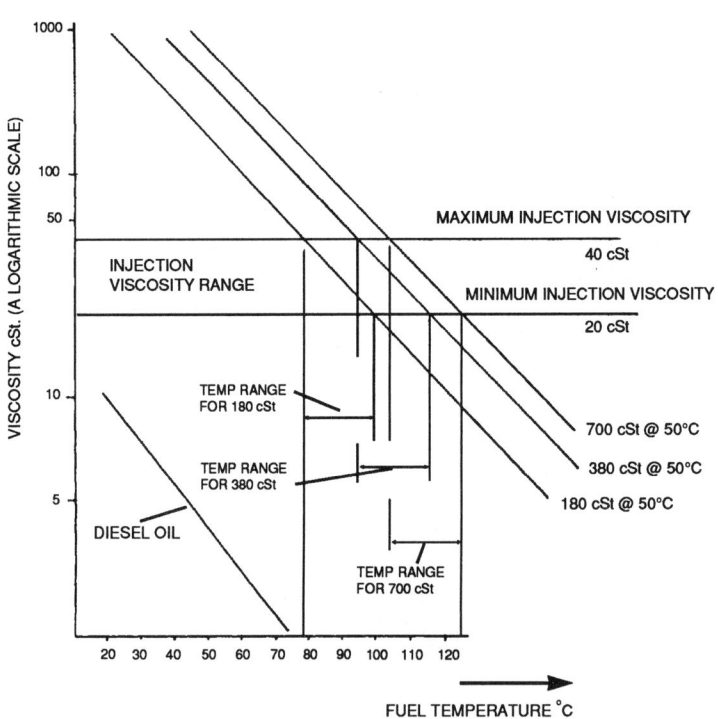

Full lift occurs quickly as the extra area of the needle seat is exposed after initial lift. The full action of lift is limited by the needle shoulder which halts against a thrust face on the holder. The injector lift pressure varies with the design but may be about 140 bar average (some designs 250 bar).

A fuel valve lift diagram for such an injector is given in Chapter 1. By removal of the spring cap the valve lift indicator needle can be assembled in the adaptor. This particular design as sketched is not cooled itself but is enclosed in an injector holder with seal (face to face) at tapered nozzle end with rubber ring at the top.

FIG 63
FUEL VALVE INJECTOR (HYDRAULIC)

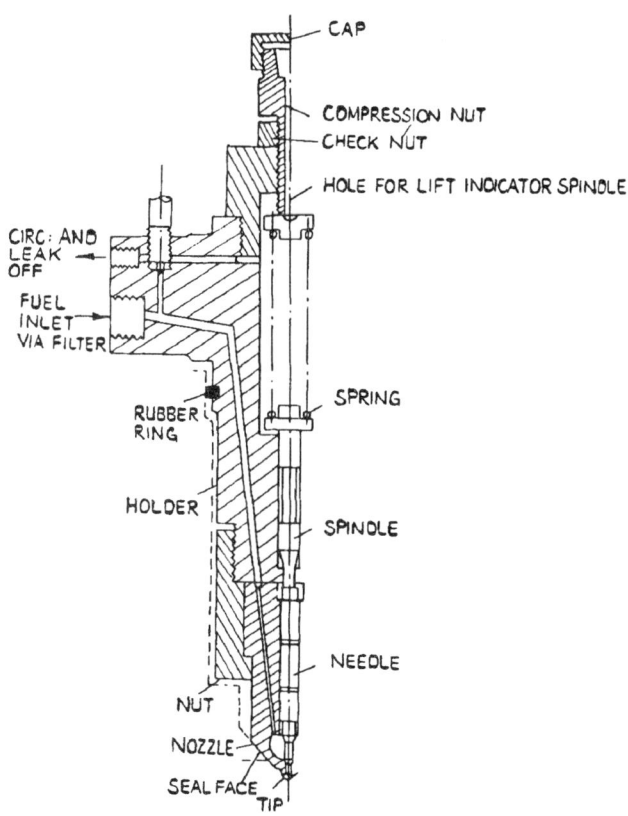

Coolant is circulated in the annular space between the injector holder and the holder itself. Direct cooling of the fuel valve as an alternative to this is easily arranged. Coolant connections on the main block would supply and return through drillings similar to that shown for fuel. The choice of oil or water for cooling depends on the engine and valve design and is also affected by the type of fuel. With hot boiler oil it is necessary to cool right to the injector tip so as to attempt to keep metal temperatures below 200°C. Hydraulic fuel valves usually have a lift of about 1 mm and the action is almost instantaneous.

There are three broad types of injection:
- Jerk injection.
- Common rail.
- Timed injection.

Jerk injection
This is by far the most common system employed on modern marine diesel engines. The fuel pressure is built up at a fuel pump in a few degrees of rotation of the cam operating the plunger. Fuel is delivered directly to spring loaded injectors which are hydraulically opened when the jerk pump plunger lift has generated sufficient fuel pressure.

Common Rail
A system in which fuel pumps deliver to a pressure main and various cylinder valves open to the main and allow fuel injection to the appropriate cylinder. Requires either mechanically operated fuel valves (*e.g.* older Doxford engines) or mechanically operated timing valves (*e.g.* modern Doxford engines) allowing connection between rail and hydraulic injector at the correct injection timing.

Timed Injection
As defined above in which the fuel pump delivers to the timing valve and thence to the spring loaded injector. No lost motion clutch is required as the cam does not drive a pump plunger but operates a valve. The cam is symmetrical with respect to engine dead centre.

Note: Many aspects of fuels are covered in Volume 8 (Chapter 2) and revision of oil tests as well as basic definitions relating to specific gravity, Conradson carbon residue, Cetane number, etc., is strongly advised.

Indicator Diagrams

Details have been given of some typical indicator diagrams showing engine faults in Chapter 1. Aspects of fuel injection faults included late and early injection (draw card), fuel valve lift diagrams, etc., as well as related details such as compression cards. Two further typical faults are as illustrated in Fig. 64.

Afterburning will show as indicated with a loss of power, increased cylinder exhaust temperature and possible discolouration of exhaust gases. Fuel restriction at filters, injectors, etc. or due to incorrect viscosity will result in a loss of power and reduced maximum pressure.

FUEL PUMPS

General

The physical energy demands of injection are great. Typical requirements include delivery of about 100ml of fuel in 1/30 second at 750 bar so as to atomise over an area of $40m^2$. A peak energy input can reach 230 kW. A short injection period at high pressure, so placed to give the desired firing pressure, is necessary. Generally pilot injection and slow injection of charge is difficult to arrange for modern turbo-charged engines.

Quantity control

The amount of fuel injected per stroke is usually accomplished by varying the effective plunger stroke of the fuel pump. This may be achieved by:

FIG 64

EFFECTS OF DEFECTIVE FUEL INJECTION

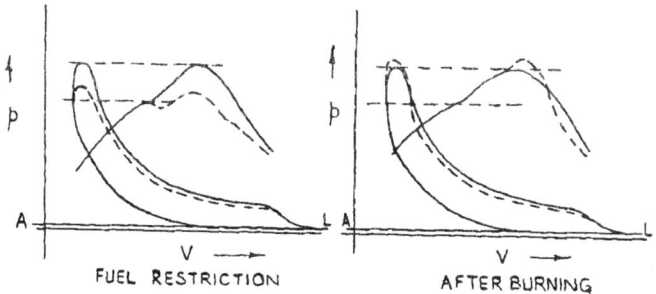

1. Varying the beginning of delivery.
2. Varying the end of delivery.
3. Varying the beginning and the end of delivery.

Control has been arranged for regulated end of effective stroke by helical groove with constant beginning of injection, as in the well known Bosch principle, described later. This method is regularly utilised for auxiliary engines and gives fuel injection early in the cycle at light load which gives higher efficiency but also leads to higher firing pressures. It has also been utilised with large direct coupled engines (*e.g.* B.& W.). Control in valve type pumps for large engines was usually with constant end and regulation of the start of injection by varying the suction valve closure (*e.g.* Sulzer and Doxford). Engine performance at low load with later injection is a compromise between economy and firing pressure. With turbo-charged engines the disadvantage of the constant end pump control is more noticeable as reduced firing pressure and efficiency is more marked at low loads due to reduced turbo-charger delivery and pressure.

Because of the limitations of varying the fuel quantity delivered by only varying the beginning of delivery Sulzer redesigned their fuel pumps to include a suction valve and spill valve. Initially the spill valve only was controlled resulting in constant beginning with variable end of delivery. Latterly, however, in the interests of fuel economy and in common with other manufacturers, both valves are controlled to give variable beginning and end of delivery.

Injection Characteristics

The diagram given in Fig. 65. illustrates some features of fuel injection based largely on Sulzer RND practice.

The fuel valve injector lift diagram is shown with lift of about 1·3 mm and injection period at full load approximately 6 degrees before to 22 degrees after. High firing pressures at full load in the high powered turbocharged range of engines can be reduced with a constant beginning method of injection. The ideal injection law corresponds to a rectangle with almost constant fuel pressure before the injector during injection. The practical curves shown show almost constant pressure at a reasonable maximum (750 bar). The last sketch of this Fig. 65. shows the effect of earlier spill at delivery for reducing load with a constant start of injection. The

FIG 65
INJECTION CHARACTERISTICS

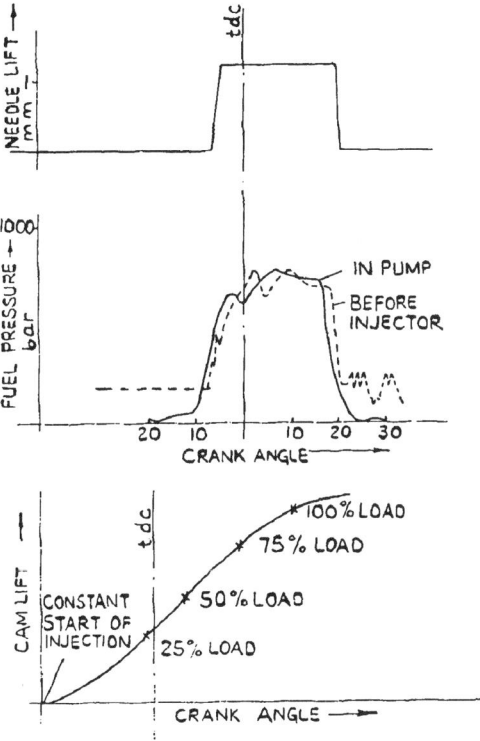

plunger at first moves very rapidly to build up pressure but is slowed during injection so giving minimum injection pressures at full load. The constant beginning of injection gives a flat combustion pressure over the full power range. At low loads the fuel pressure is higher than is usually the case because the plunger is delivering more in its maximum speed range, this gives better atomisation.

Variable Injection Timing (VIT)
The previous section dealt with only variable and with constant beginning of injection. It is modern practice for economy considerations to now vary both the beginning and end of injection. With constant beginning of injection the maximum firing

FIG 66
FUEL SAVINGS AVAILABLE BY UTILISING VARIABLE INJECTION TIMING (VIT)

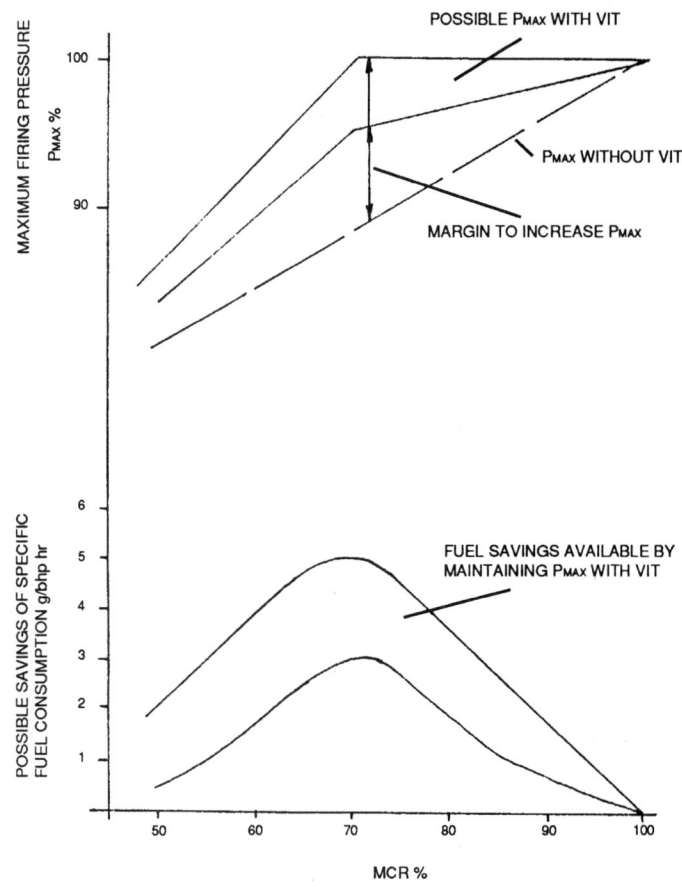

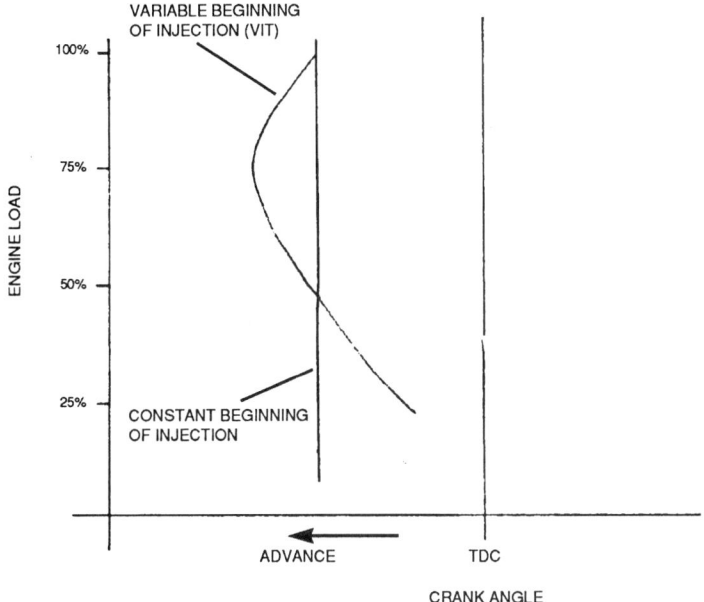

**FIG 67
INJECTION TIMING VARIATION
WITH ENGINE LOAD**

pressure of the engine will fall almost linearly as the power of the engine is reduced. The b.m.e.p of the engine, however, reduces at a slower rate and since the thermal efficiency of the engine varies as the ratio of P_{MAX}/P_{MEP} then a reduction of firing pressure will result in a reduction of thermal efficiency of the engine. In order that the thermal efficiency and hence the specific fuel consumption can be maintained at optimum it is therefore necessary to maintain maximum firing pressures as the engine load is reduced. This is accomplished by advancing the timing of the fuel injection as the engine load is reduced Fig. 66. The advancement of the injection timing continues until about 65%-70%, thereafter the injection is retarded. Fig. 67. It can be seen that at about 25% engine load injection is retarded in relation to full load timing. This will reduce the "diesel knock" at low engine loads that is sometimes experienced on engines without variable injection timing [VIT].

FIG 68
BOSCH TYPE FUEL PUMP OPERATION

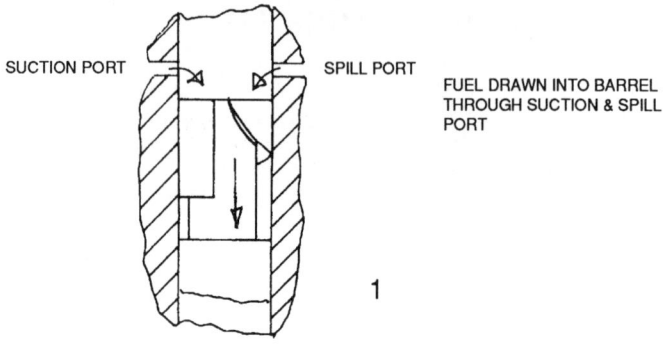

1

FUEL DRAWN INTO BARREL THROUGH SUCTION & SPILL PORT

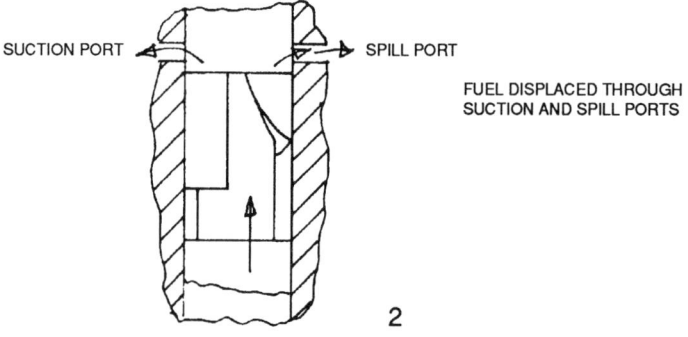

2

FUEL DISPLACED THROUGH SUCTION AND SPILL PORTS

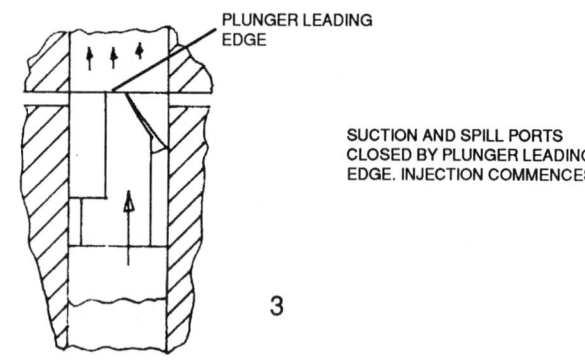

3

SUCTION AND SPILL PORTS CLOSED BY PLUNGER LEADING EDGE. INJECTION COMMENCES

FIG 68
LONGITUDINAL GROOVE & PLUNGER CONTROL

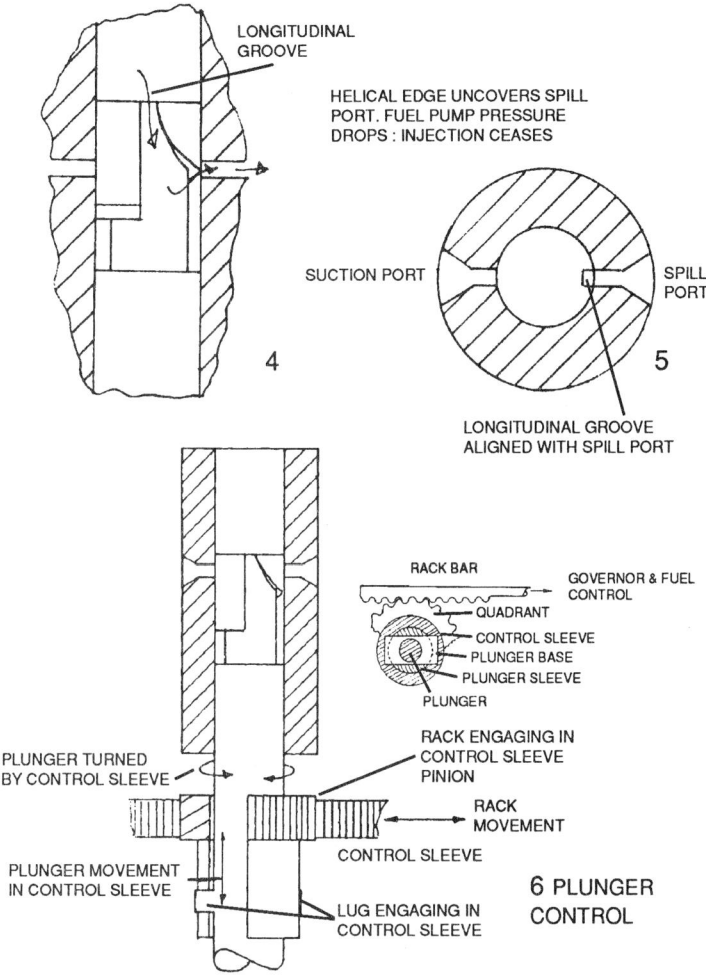

Bosch Jerk Pump Principle
Fig. 68. shows a Bosch type fuel pump set at approximately 75% load.

The sketch numbered 1 show the plunger moving down. The pressure in the barrel falls and as the suction and spill ports open to the fuel rail the fuel flows into the barrel.

In Sketch 2 the plunger is moving upwards. The fuel is displaced from the barrel through the spill and suction ports. This displacement will continue until the plunger completely covers both ports.

Sketch 3 shows the plunger continuing to move upwards and just covering the spill and suction ports. This is the effective beginning of delivery and any further upward movement of the plunger will pressurise the fuel and open the fuel valve injecting fuel to the engine.

Sketch 4. The plunger continues to move upwards. Injection continues until the point when the helical edge of the groove on the plunger uncovers the spill port. The high pressure in the barrel is immediately connected to the low pressure of the fuel suction. There is no longer sufficient pressure to keep the fuel valve needle open and injection ceases.

Sketch 5 shows the plunger turned by means of the rack and pinion so that the longitudinal groove of the plunger is aligned with the spill port. In this position the plunger is unable to deliver any fuel since the spill port does not close during the cycle.

Sketch 6 shows a sectional plan view to illustrate how the plunger can be rotated so as to vary helix height relative to the spill port. The plunger base is slotted into a control sleeve which is rotated by quadrant and rack bar.

With both valve and helical spill designs there can be problems of fuel cavitation due to very high velocities. Velocities near 200 m/s can create low pressure vapour bubbles if pressure drops below the vapour pressure. These bubbles can subsequently collapse during pressure changes which results in shock waves and erosion attack as well as possible fatigue failure. A spring loaded piston and orifice design can absorb and damp out fluctuation. Götaverken, Stork, MAN and B & W. engines utilise a form of the above pump.

The adjustment of injection timing is carried out on Bosch type fuel pumps by varying the relative height of plunger and suction/spill ports in the barrel. This may be accomplished in a number of ways:

1. Adjusting the height of plunger relative to the barrel. Fig 69a.

2. Adjusting the height of the barrel relative to the plunger. Fig 69b.

FIG 69
ADJUSTMENT OF FUEL INJECTION TIMING BY VARYING RELATIVE POSITION OF PLUNGER AND BARREL

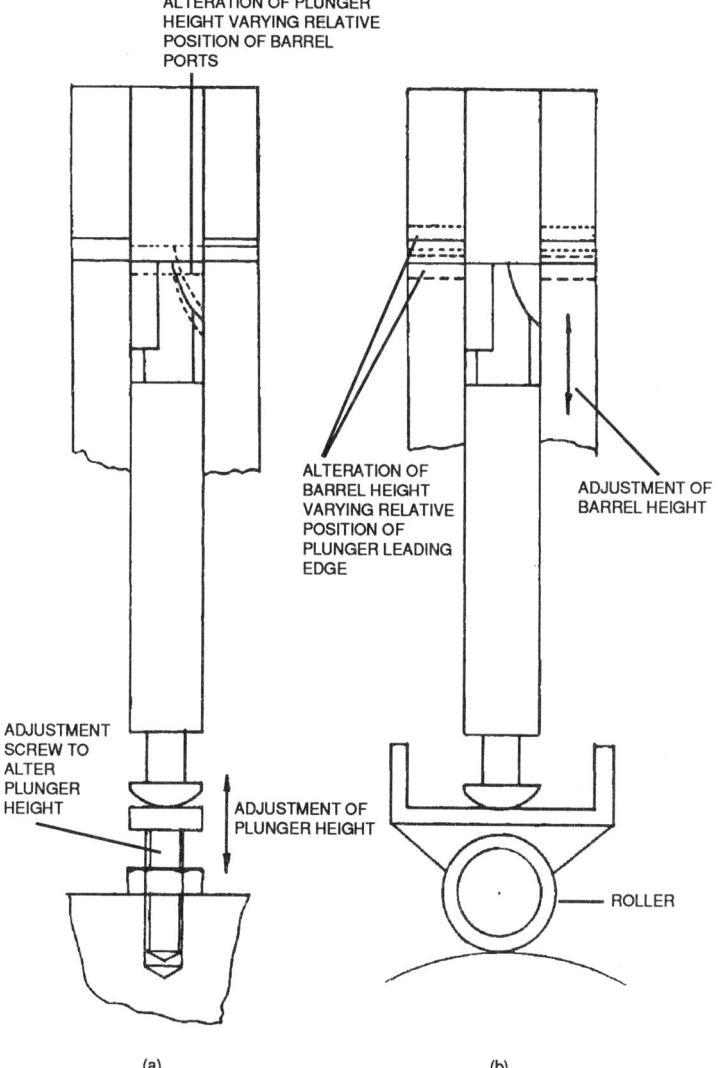

Adjustment of plunger height can be accomplished in some installations by adjusting the cam follower. Lowering the plunger has the effect of retarding the injection. Raising the plunger advances the injection.

In earlier B & W. designs adjustment of injection timing was carried out by raising the fuel pump barrel in relation to the plunger. In this design the fuel pump top flange has an external threaded portion projecting towards the barrel Fig. 70. This thread matches with the external thread of the adjusting ring. The adjusting ring has external gear teeth cut on its upper part which are engaged by the adjusting pinion. To adjust the injection timing the pinch bolts are released and the adjusting pinion turned to either raise or lower the barrel. Lowering the barrel will advance the injection timing while raising the barrel will retard the injection timing.

Jerk Fuel Pump [valve type fuel pump] Detail

The detail in Fig. 71. is based on an earlier Sulzer design. The fuel-pump delivery is controlled by suction and spill valves.

Sketch 1 shows the plunger moving down, the suction valve open and fuel-oil being drawn into the barrel.

Sketch 2 shows the plunger moving upwards and fuel being displaced through the still open suction valve.

Sketch 3 shows the delivery commencing as the suction valve closes.

Sketch 4. As the plunger continues to move upwards injection ceases when the spill valve opens.

From Fig. 71. it can be seen that the valve type fuel pump design lends itself quite readily to control on suction valve and spill valve. However, scroll type, or Bosch fuel pumps require a different method to control the beginning and the end of injection. The beginning of injection is carried out by adjusting the height of the fuel pump barrel. Referring to Fig. 72. the pump barrel has an outside threaded lower portion which engages into the timing guide operated by a toothed rack. Movement of the rack causes the pump barrel to move vertically up or down relative to the pump plunger. In this way the moment the plunger covers the spill port, and so commences injection, can be adjusted while the engine is in operation. The pump barrel is prevented from turning by a locating

FIG 70
ADJUSTMENT OF INJECTION TIMING BY RAISING OR LOWERING FUEL PUMP BARREL

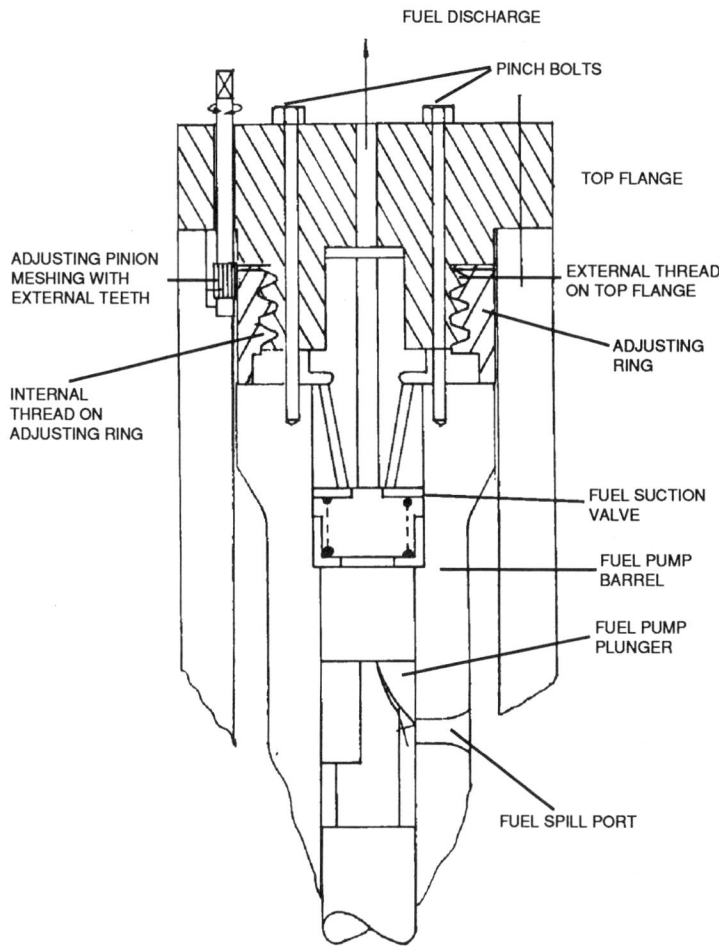

FIG 71
SULZER VALVE TYPE FUEL PUMP

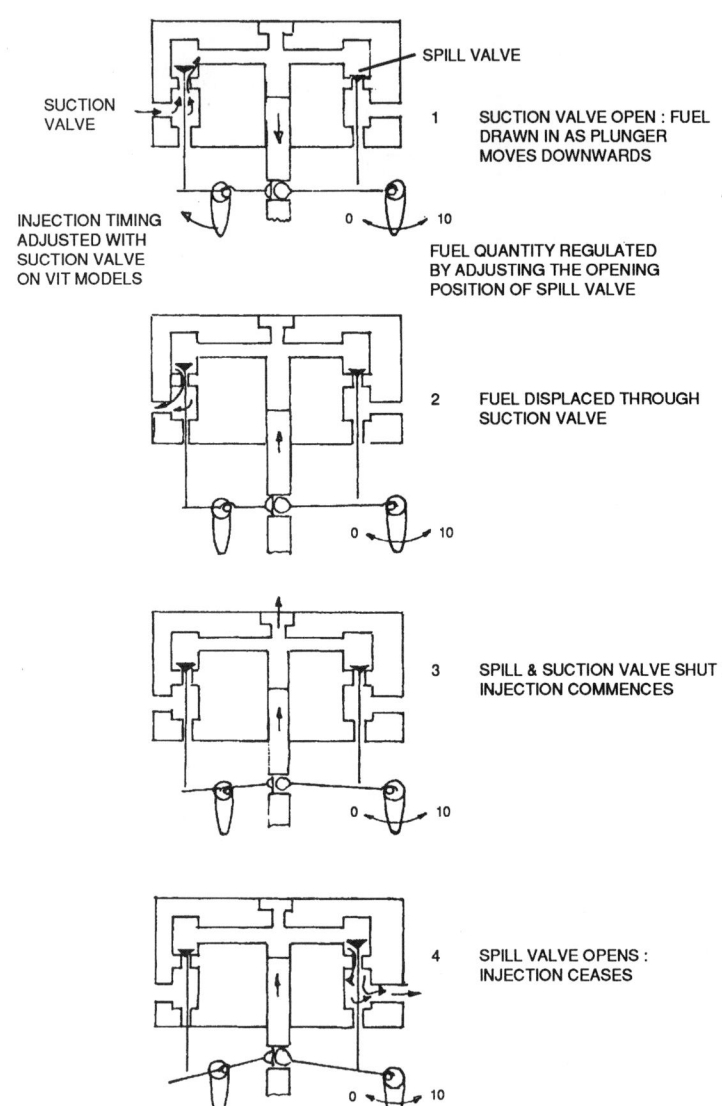

FIG 72
VIT BOSCH TYPE FUEL PUMP

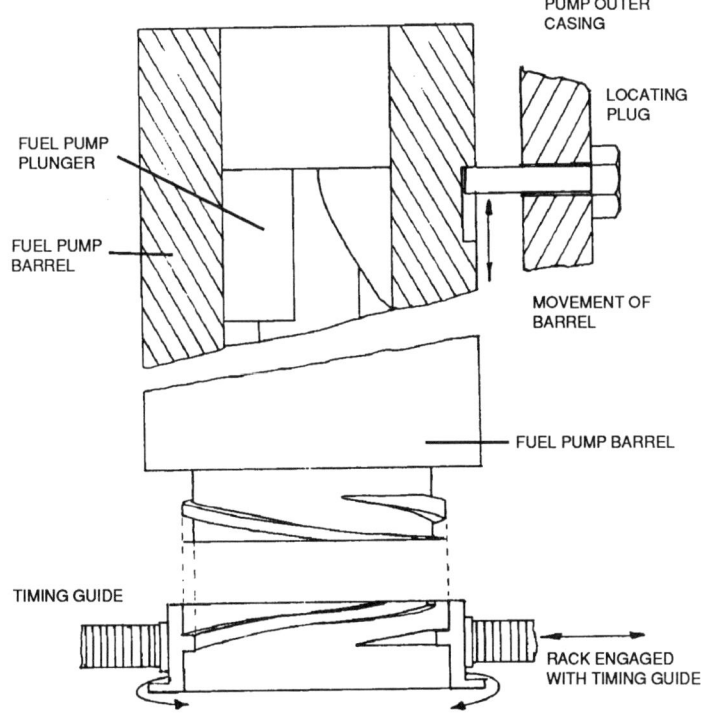

plug. The end of injection, and therefore the quantity of fuel delivered, is regulated by rotating the plunger which varies the position of the helix edge relative to the spill port.

Two Stage Fuel Injection
Efforts are continually being made to improve the reliability and economy of medium speed engines which operate on heavier grades of fuel. One way of achieving this is by having a carefully controlled, reliable combustion process. This requires good atomisation with short injection periods but this results in high injection rates at high engine loads with commensurate high rates at intermediate and low loads where increased ignition delays may be experienced. Indeed, research has shown that, at low loads, the injection process may be completed before ignition commences. Because the fuel is well mixed with the air, the combustion is very intense and almost instantaneous when ignition does eventually occur. This uncontrolled energy release will cause "diesel knock" and possibly destructive thermal and mechanical stresses.

As a solution to this problem Wärtsilä have developed a two-stage injection process in which the fuel injected during the pilot stage is constant and independent of engine load. The quantity of fuel injected during the pilot phase is set at about 2·6% of the maximum continuous rating [MCR], this is marginally less than the amount required to compensate for frictional loses when the engine is idling. The pilot fuel is injected in advance of the main injection phase but the quantity involved is too small to damage the combustion chamber components. Fig. 73. The injection and ignition of the pilot fuel minimises the ignition delay because it raises the temperature of the combustion air. The fuel injected during the main stage enters a favourable environment with combustion commencing as soon as the first fuel droplets enter the combustion chamber. This eliminates the possibility of unburnt fuel being stored in the combustion chamber and hence the destructive uncontrolled release of energy.

Both fuel valves are supplied by the same fuel pump. The fuel pump, however, has two plungers to supply pilot and main injection. The main plunger of this fuel pump is of the conventional scroll, or Bosch type. The pilot plunger is positioned above the main plunger, but since the quantity of fuel injected in

FIG 73
PILOT AND MAIN INJECTION AGAINST CRANK ANGLE

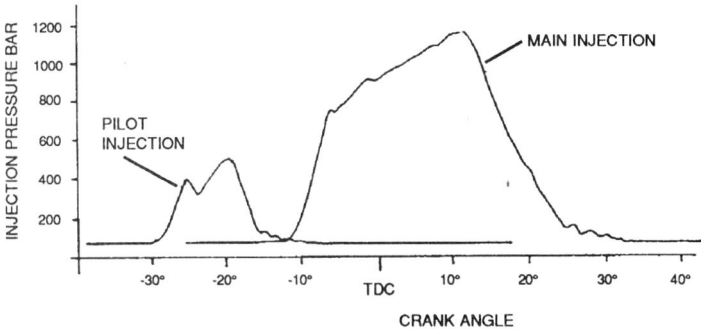

the pilot stage is constant this plunger has no helix. As the pilot plunger covers the suction/spill port injection commences and ceases as the lower edge of the plunger uncovers the suction/spill port. Fig. 74.

This arrangement has the advantage that the emission of NOx [oxides of nitrogen] are reduced and allows the use of low Cetane fuels.

Fuel Systems

In recent years the quality of the fuel available to the marine industry has deteriorated. This had led inevitably, not only to problems of combustion, but also to problems of storage.

To alleviate some of these problems the entire fuel system should be carefully designed and this should commence with the bunkering system. Modern fuels tend to be high viscosity and may have a high pour point so it is important that at the completion of bunkering that the fuel drains freely into the bunker tanks. If loading bunkers in cold climates it may be necessary to include insulation on the exposed bunker lines. An indication of a high pour point fuel may be a high loading temperature: the supplier ensuring that the fuel is easily pumped on board. If a waxy fuel is suspected then a pour point test should be carried out.

FIG 74
TWO STAGE FUEL PUMP

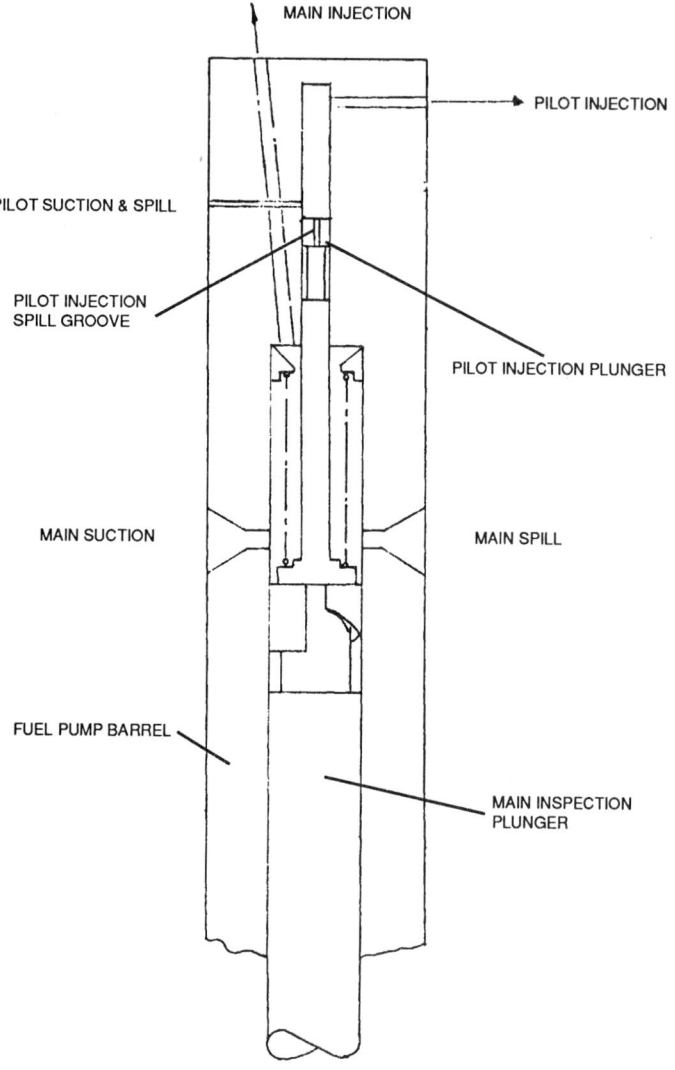

FIG 75
POSITION OF MAIN AND PILOT INJECTION IN CYLINDER

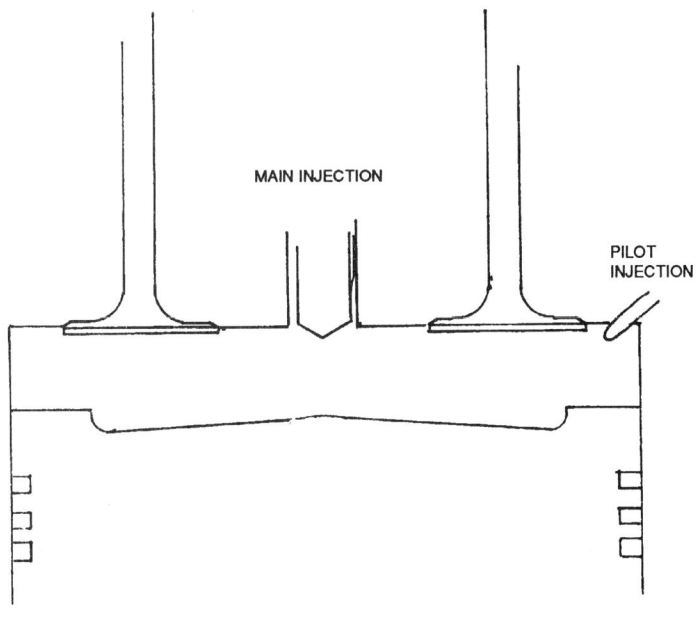

Due to the problems associated with incompatibility fuels from different sources should not be mixed. The importance of the segregation of fuels from different sources cannot be overstated and should be achieved, wherever possible, by transferring remaining fuel into smaller tanks prior to bunkering in order that the total quantity of fuel loaded can be received in empty tanks. Even if a vessel is equipped with adequate storage to ensure segregation mixing may occur in the settling and service tanks when fuels are changed over. If compatibility problems are suspected then fuel change-overs should be accomplished by running down the settling tank before admitting the next, possibly, incompatible fuel. Two service and two settling tanks would be the ideal situation, since fuel change-overs would be accomplished with the absolute minimum mixing. Designers and shipowners may wish to consider this.

The temperature of the stored fuel must be monitored to ensure that it does not approach its pour point. This is important, especially when fuel is stored in double bottoms, since it is not uncommon for fuels to have a pour point of 25°C and so become unpumpable even at temperatures that can be considered temperate.

The heating capacity of the fuel system, that is, tank heating, trace heating and system heating should be able to deal with the viscosity of any fuel the vessel is likely to encounter.

Tank heating must be able to maintain temperatures above the maximum likely pour point.

Settling and service tank heating should maximise the settling of water and solid matter from the fuel and maintain the correct post-purification temperature. This is important since Department of Transport regulations require that fuel is stored at, or below 60°C. However, the purification and clarification temperatures of high viscosity fuels may be substantially higher than this; 100°C for example. To comply with this regulation a post-purifier FUEL COOLER may be required.

When operating with high viscosity fuels it is necessary to employ high rates of heat transfer during fuel heating. This could cause thermal cracking of the fuel leading to carbon deposits on the heating surfaces resulting in reduced heating capacity. To maintain optimum heat transfer and heating steam consumption there should be a facility to enable the oil side of the heater to be cleaned periodically by circulating with a proprietary carbon remover. A Typical Fuel System is shown in Fig. 76.

Fuel management
Many fuel related problems that have occurred on ships in recent years could have been avoided if an effective "on board fuel management policy" had been adopted. Such a policy would include.

1. Representative samples of fuel, in addition to the suppliers sample, taken at loading. These should be sealed, clearly labelled and retained on board for 3 months after the fuel has been consumed.

2. Segregation of fuels from different sources by loading into empty tanks. This may involve the transfer of remaining fuels into smaller tanks prior loading.

3. Draining bunker lines at the completion of loading. Closing all bunker valves when this is accomplished.

4. Maintaining storage temperatures at least 5°C above the pour point.

5. Sending a representative fuel sample for analysis and taking the appropriate action upon receipt of the results.

6. Draining water from tanks at regular intervals.

7. Monitoring fuel consumption against fuel remaining on board. This should be achieved by daily measurement of all fuel tanks: Dipping tanks probably provides the most accurate results.

Fuel analysis may reveal:
1. *Low flash point:* D.Tp. regulations require a flash point above 65°C. If FP below this is encountered the owners and classification society should be informed. A lower flash point fuel will render the vessel "out of class". The addition of a higher flash point fuel will not raise the flash point of the original stock. To avoid the generation of a flammable vapour heating temperatures should be regulated carefully.

2. *High sulphur:* May result in acids forming which may deplete the TBN of cylinder/system oil. Engine may require a higher TBN oil.

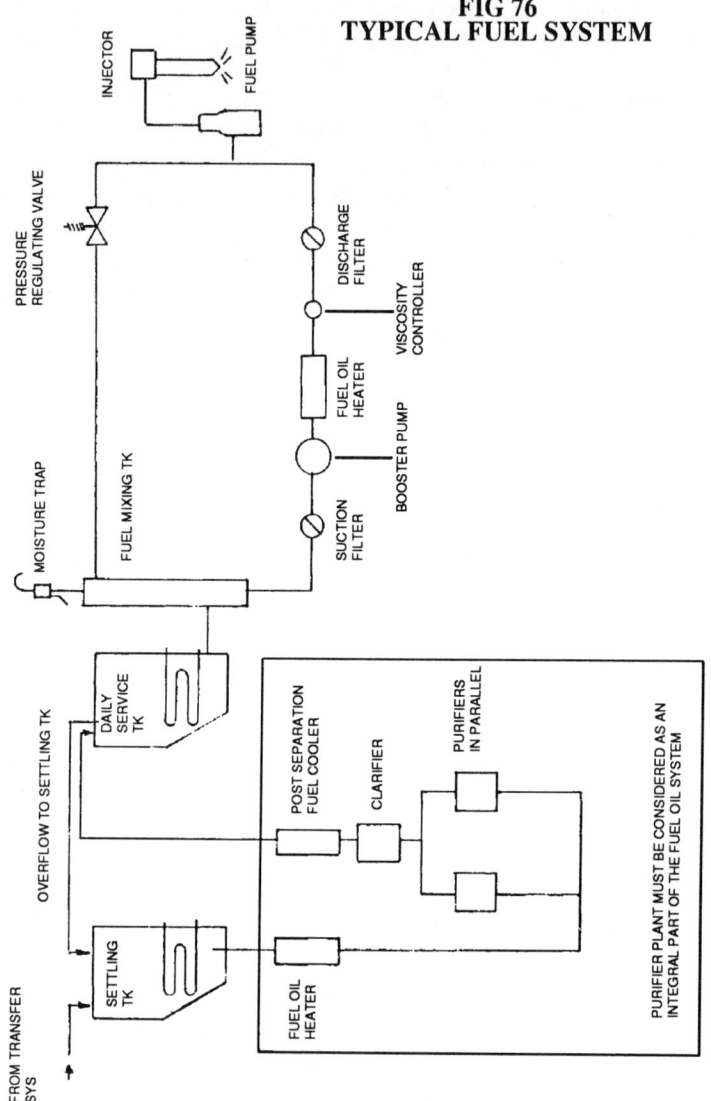

FIG 76
TYPICAL FUEL SYSTEM

3. *High water content:* May separate when heated. May, however, form a stable emulsion which is difficult to separate without the addition of emulsion breaking chemicals. If water contamination is salt water, not uncommon in the marine environment, serious problems associated with sodium-vanadium corrosion and turbo-charger fouling may be experienced. Water contamination also introduces the risk of bacteria into the fuel. Bacterial growth can occur at the oil/water interface which if allowed to proliferate can cause blockage of filters and fuel system. The problem of bacterial, or microbial attack, is greater in fuel which is unheated, especially diesel oil, since the temperatures involved when heating high viscosity fuels will pasteurise the fuel and thus kill off bacteria. Since prevention is better than cure draining the water from the oil is probably the best course of action.

4. *High vanadium:* May cause high temperature corrosion. The use of an ash modifying chemical additive to maintain the vanadium oxides in a molten state will prevent adhesion to high temperature components.

5. *Instability and incompatibility:* Instability refers to tendency of the fuel to produce a sludge by itself. Incompatibility is the tendency of the fuel to produce a sludge when blended with other fuels. These sludges form when the asphaltene content of the fuel can no longer stay in solution and so precipitates out, sometimes at a prodigious rate. The deposited sludge blocks tank suctions, filters, and pipes and quickly chokes purifiers. In engines the blockage of injector nozzles, late burning and coking can result in damage to pistons, rings and liners.

6. *High aluminium content:* This contamination is a result of carry over of "catalytic fines" from the refining process of the initial feedstock oil. These "fines" are an aluminium compound ranging in size from 5 mm to 50 mm and are extremely abrasive. Very low levels of aluminium indicate the presence of catalytic fines in the fuel which, if used, will lead to high levels of abrasive wear in the fuel system, piston, rings and liner in an extremely short period of time. 30 ppm of aluminium is generally considered as the maximum allowable level in fuel oil bunkers before purification. Because of the small size of these compounds they are difficult to remove completely by centrifuge. The purification

plant, in correct operation, will reduce the aluminium content to about 10 ppm before it is used in the engine. It has been found that if the aluminium content is above 30 ppm difficulties will be experienced attaining a safe level of 10 ppm after purification.

CHAPTER 4

SCAVENGING AND SUPERCHARGING

If maximum performance and economy, etc. are to be maintained it is essential during the gas exchange process that the cylinder is completely purged of residual gases at completion of exhaust and a fresh charge of air is introduce into the cylinder for the following compression stroke. In the case of 4-stroke engines this is easily carried out by careful timing of inlet and exhaust valves where, because of the time required to fully open the valves from the closed position and conversely to return to the closed position from fully open, it becomes necessary for opening and closing to begin before and after dead centre positions if maximum gas flow is to be ensured during exhaust and induction periods. Typical timing diagrams are shown in Fig. 77. for both normally aspirated and pressure charged 4-stroke engine types. Crank angle available for exhaust and induction with normally aspirated engines is seen to be of the order 420° to 450° with a valve overlap of 40° to 60° depending upon precise timing – with more modern pressure-charged engines this increases to around 140° of valve overlap. Basic object of overlap, *i.e.* exhaust and inlet valves open together, is to assist in final removal of remnant exhaust gases from cylinder so that contamination of charge air is minimal. The extension of overlap in the case of pressure-charged engines serves to: (1) further increase this scavenge effect, (2) provide a pronounced cooling effect which either reduces or maintains mean cycle temperature to within acceptable limits even though loading may be considerably increased. Consequent upon (2) it becomes clear that thermal stressing of engine parts is relieved and with exhaust gas turbocharger operation prolonged running at excessively high temperatures is avoided. This latter would have an adverse effect on materials used in turbocharger construction and could also contribute toward increased contamination.

In some cases an apparent anomaly exists between the temperature of exhaust gas leaving the cylinder and the temperature at inlet to turbocharger, being as much as 90° higher. This is

FIG 77
TYPICAL TIMING DIAGRAMS

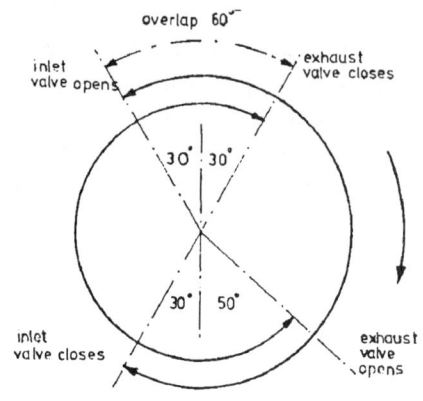

NATURALLY-ASPIRATED 4-STROKE

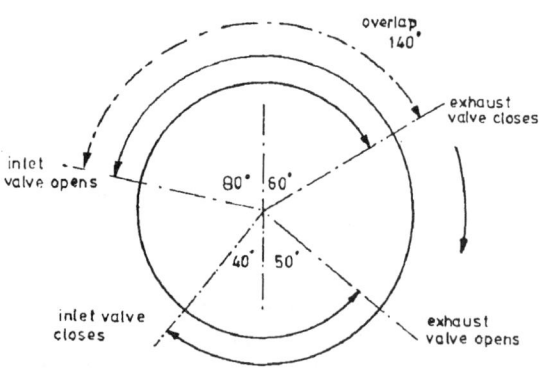

PRESSURE-CHARGED 4-STROKE

partially explained by the fact that over the latter part of the gas exchange process the relatively cold scavenge air will have a depressing effect on the temperature indicated at cylinder outlet which will tend to indicate a mean value over the cyclic exchange.

More probably the increase may be largely attribute to change of kinetic energy into heat energy and an approximately adiabatic compression of the gas column between cylinder and turbine inlet.

2-Stroke Cycle Engines

With only one revolution in which to complete the cycle the time available for clearing the cylinder of residual exhaust gases and recharging with a fresh air supply is very much reduced. Of necessity the gas exchange process is carried out around b.d.c. where the positive displacement effects of the piston cannot be exploited as is the case with the 4-stroke cycle. The total angular movement seldom exceeding 140° compared to well in excess of 400° with 4-stroke operation gives some indication of the need for high efficiency scavenging processes if cylinder charge is not to suffer progressive contamination and subsequent loss of performance with increased temperature and thermal loading. Prior to the introduction of turbocharging to 2-stroke machinery this necessitated a low degree of pressure-charging of 1·1 to 1·2 bar to ensure adequacy of the gas exchange process.

The scavenging of 2-stroke engines is generally classified as: (1) uniflow or longitudinal scavenge, (2) loop and cross scavenge. Fig. 78. Because of the simplicity of the arrangement uniflow scavenge as employed in poppet-valve or opposed piston type engines is generally considered as being the most efficient. In this case charge air is admitted through ports at the lower end of the cylinder and as it sweeps upwards toward the exhaust discharge areas, almost complete evacuation of residual gases is obtained. By suitable design of the scavenge ports or the provision of special air deflectors the incoming charge air can be given a swirling motion which intensifies the purging effect and also promotes the degree of turbulence within the charge which is required for good combustion when fuel injection takes place.

Cross and loop scavenge have both exhaust and scavenge ports arranged around the periphery of the lower end of the liner and in so doing eliminate the need for cylinder head exhaust valves etc and their attendant operating gear. This considerably simplifies engine construction and can lead to a reduction of maintenance. Because of simplified cylinder head construction the cylinder combustion space can be designed for optimum combustion conditions. Generally however, the scavenging efficiency is somewhat lower than with the uniflow system due to the more complex gas-air interchange and the possibility of charge air passing straight to exhaust with little or no scavenging effect. Careful attention to port design does however considerably reduce this problem.

The gas exchange process itself may be divided into three separate phases: (1) blowdown, (2) scavenge and (3) post-scavenge.

During blowdown the exhaust gases are expelled rapidly – the process being assisted by amply dimensioned ports or valves arranged to open rapidly. At the end of this blowdown period when the scavenge ports begin to uncover, the cylinder pressure should be at or below charge air pressure so that the scavenge process which follows, effectively sweeps out the remaining residual gases. With scavenge ports closed the post scavenge period completing the gas exchange process should ensure that exhaust discharge areas close as quickly as possible to prevent undue loss of charge air so that the trapped air at beginning of compression has the highest possible density. Although some loss of charge air is unavoidable it should be borne in mind that the air supply is considerably in excess of that required for combustion and the cooling effect of the air passing through the system has the result of keeping mean cycle temperatures down so that service conditions are less exacting.

The increased cylinder pressures encountered with modern turbocharged machinery may result in exhaust opening being advanced so that sufficient time is given for cylinder pressure to fall to or below charge air pressure when the scavenge ports uncover. A complementary aspect of earlier opening to exhaust is the increased pulse energy obtainable from the exhaust gas which can be utilised to improve turbocharger performance. In many cases this is the main criterion which influences exhaust opening, since the loss of expansive working is more than offset by the gain in turbocharger output.

Obviously in the case of reversing engines there may be some slight penalty incurred if prolonged operation in the astern direction is considered. Fig. 79. shows the timing for some of the present generation of direct drive slow speed diesels.

FIG 78
SCAVENGING OF 2-STROKE ENGINES

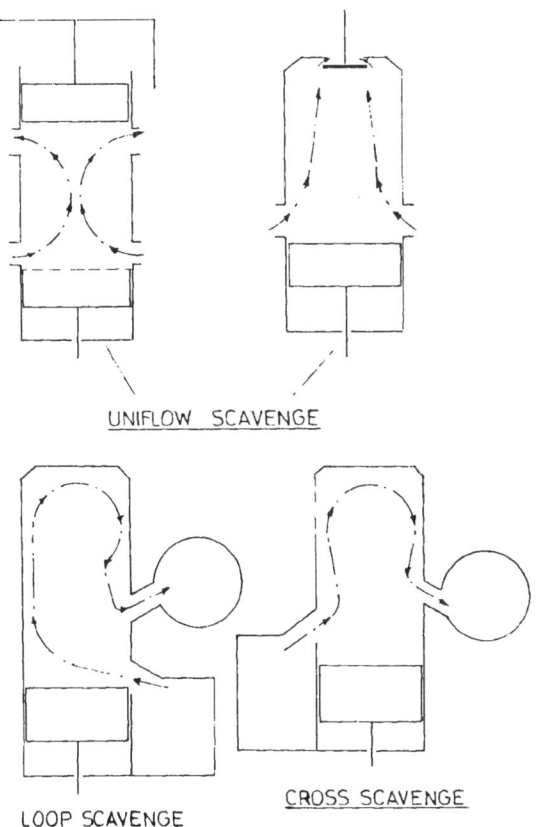

UNIFLOW SCAVENGE

LOOP SCAVENGE   CROSS SCAVENGE

PRESSURE CHARGING

By increasing the density of the air charge in the cylinder at the beginning of compression a corresponding greater mass of fuel can be burned giving a substantial increase in power developed. The degree of pressure charging required, which determines the increase in air density, is achieved by the use of free running turbochargers which are driven by the exhaust gases expelled from

FIG 79
TIMING FOR SOME DIRECT DRIVE SLOW SPEED DIESELS

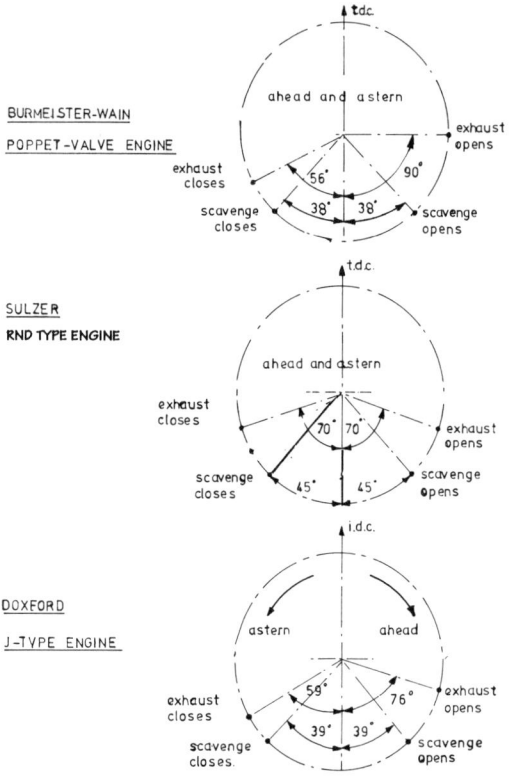

the main engine. About 20% of the energy available in the exhaust gas is utilised in this way. In the past it was usual practice to employ some form of scavenge assistance either in series or parallel with the turbochargers. This was accomplished by engine driven reciprocating scavenge pumps, under piston effect or independently driven auxillary blowers. Only the under-piston effect and auxillary blowers are used to any significant degree in modern practice.

FIG 80
PRESSURE-CHARGING

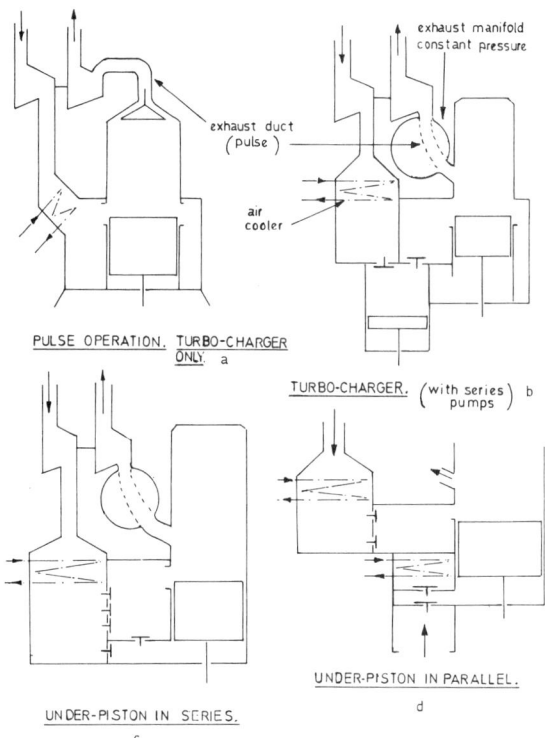

Fig. 80. (b) and (c). turbocharger provides charge air at 70 to 95% of required pressure with under-piston effect or series pump making up the balance. Slight increase in temperature of air delivered to engine since air cooling is carried out after the turbocharger only.

Fig. 80. (d). With parallel operation air supply to engine is increased by air delivery from pumps with proportionate increase in output resulting in greater exhaust gas supply to turbocharger and improved turbocharger performance.

The advantages of pressure-charging may be summed up as: (1) substantial increase in power for a given speed and size; (2) better mass power ratio, *i.e.* reduced engine mass for given output; (3) improved mechanical efficiency with reduction in specific fuel consumption; (4) reduction in cost per unit of power developed; (5) the increase in air supply has a considerable cooling effect leading to less exacting working conditions and improved reliability. Because of increasing power output and fuel economy diesel plant is now almost universally chosen for applications once dominated by steam turbine plant.

Constant Pressure and Pulse Operation
In general the manner in which the energy of the exhaust gases is utilised to drive the turbocharger may be ascribed to (1) the pulse system of operation and (2) constant pressure operation.

Pulse Operation
This makes full use of the higher pressures and temperatures of the exhaust gas during the blow-down period and with rapidly opening exhaust valves or ports the gases leave the cylinder at high velocity as pressure energy is converted into kinetic energy to create a pressure wave or pulse in the exhaust lead to the turbocharger. For pulse operation it is essential that exhaust leads from cylinder to turbine entry are short and direct without unnecessary bends so that volume is kept to a minimum. This ensures optimum use of available pulse energy and avoids the substantial losses that could otherwise occur with a corresponding reduction in turbo- charger performance. Of necessity, exhaust ducting must be arranged so that the gas-exchange processes of cylinders serving the same turbocharger do not interfere with each other to cause pressure disturbances that would affect purging and recharging with an adverse effect upon engine performance. With 2-stroke engines the optimum arrangement is three cylinder grouping with 120° phasing which gives up to 10% better utilisation of available energy than cylinder groupings other than multiples of three. Due to the small volume of the exhaust ducting and direct leading of exhaust to turbine inlet the pulse system is highly responsive to changing engine conditions giving good performance at all speeds. Theoretically turbocharging on the pulse system does not require

any form of scavenge assistance at low speeds or when starting. In practice however the use of an auxiliary blower or some other means of assistance is employed to ensure optimum conditions and good acceleration from rest.

Constant Pressure Operation
In this system the exhaust gases are discharged from the engine into a common manifold or receiver where the pulse energy is largely dissipated. Although the pulse energy is lost, the gas supply to the turbine is at almost constant pressure so that optimum design conditions prevail since, under normal conditions, gas flow will be steady rather than intermittent. Further, as engine ratings increase, the constant pressure energy contained in the exhaust gas becomes increasingly dominant so that sacrifice of pulse energy in a large volume receiver is of less consequence. Fig. 81. shows the results of tests carried out on a Sulzer type engine which indicates that up to b.m.e.p. of around 7 bar the advantage lies with the pulse system but as b.m.e.p. increases beyond this figure the constant pressure system becomes more efficient giving greater air throughout and some slight reduction in the fuel rate.

Due to the much larger volume of the exhaust system associated with constant pressure operation the release of exhaust gas is rapid and earlier opening to exhaust is generally only necessary to ensure cylinder pressure has fallen to or below the charge air pressure

**FIG 81
AIR DELIVERY**

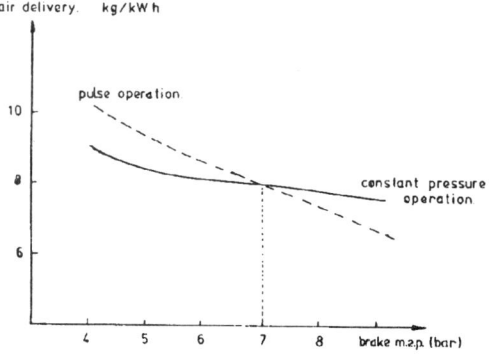

when the scavenge ports begin to uncover. With a possible reduction in exhaust lead expansive working can be increased which is a further contributory factor in reducing the fuel rate. A major drawback to constant pressure operation is that the large capacity of the exhaust system gives poor response at the turbocharger to changing engine conditions with the energy supply at slow speeds being insufficient to maintain turbocharger performance at a level consistent with efficient engine operation. Some form of scavenge assistance such as under-piston scavenging is often utilised. To offset this however the number of

**FIG 82
RND 90 SUPERCHARGING ARRANGEMENT**

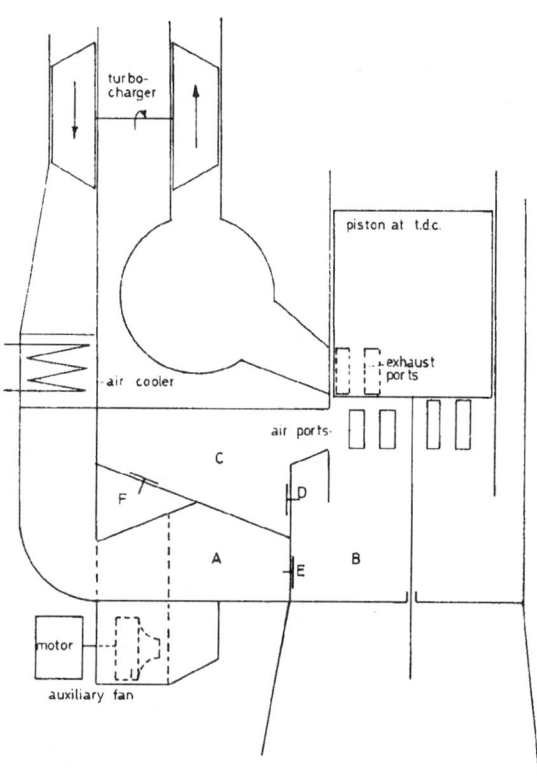

turbochargers required as compared to pulse operation can be reduced, a greater flexibility exists in the case of turbocharger location and exhaust arrangement and no de-rating of engine need be considered for cylinder groupings other than multiples of three. For this reason most large slow speed 2-stroke engines tend to be of the constant pressure configuration.

Fig. 82. shows the diagrammatic arrangement of the Sulzer RND engine which operates with constant pressure supercharge. In normal operation air is drawn into under-piston space B from common receiver A and compressed on downstroke of piston to be delivered into space C so that when scavenge ports uncover purging is initiated with a strong pressure pulse. As soon as pressure in spaces B/C falls to common receiver pressure in A scavenge continues at normal charge air pressure. For part load operation the auxiliary fan is arranged to cut in when charging pressure falls below a pre-set value. Air is drawn from space A and delivered into space F and this together with under piston effect ensures good combustion and trouble-free operation under transient conditions.

AIR COOLING

During compression of the air at the turbo-blower, which is fundamentally adiabatic, the temperature may increase by some 60-70°C with a corresponding reduction in density. This means that the air must be passed through a cooler on its passage to the engine in order to reduce its temperature and restore the density of the charge air to optimum conditions. Correct functioning of the cooler is therefore extremely important in relation to efficient engine operation. Any fouling which occurs will reduce heat transfer from air to cooling medium and it is estimated the 1°C rise in temperature of air delivered to the engine will increase exhaust temperature by 2°C. Reduction in air pressure at cooler outlet due to increased resistance is also a direct result of fouling. It is therefore imperative that air coolers are kept in a clean condition. It is preferable that this is accomplished on an ongoing basis rather than changing a dirty cooler since progressive fouling will have an adverse effect on engine performance. Ongoing cleaning can be carried out by spraying with a commercial air cooler cleaning solvent. Under conditions of high humidity precipitation at the cooler may be copious. Carry over of this water

FIG 83
WATER SEPARATOR

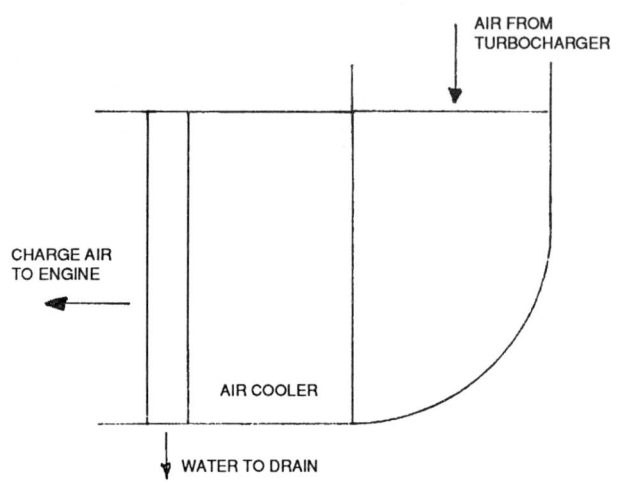

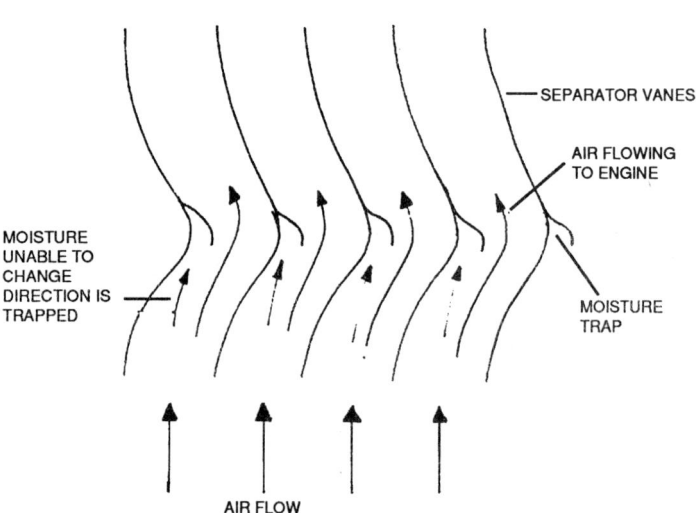

to the engine can have a number of detrimental effects. Water contamination of cylinder lubricating oil may reduce its viscosity and hence its ability to withstand the imposed loads leading to increased cylinder and piston ring wear. Water contamination may also lead to corrosion of engine components. To prevent the carry-over of water a water separator is fitted. Fig. 83. shows a water separator fitted on the outlet side of an air cooler. This separator utilises the difference in the mass of water and air. As the moist air flows into the vanes its direction is changed. Because of its lower mass the air is able to change direction easily to flow around the vanes. The water, however, because of its greater mass and, therefore momentum, is not able to change direction so easily and flows into the water trap to be removed at the drain. The water separator should also be sprayed with cleaning solvent when cleaning the air cooler. It must be noted that the vapour given off by cleaning solvents is harmful and by spraying into air coolers may contaminate the atmosphere throughout the engine. The air coolers should not be cleaned when personnel are working within the engine.

TURBOCHARGERS

These are essentially a single stage axial flow turbine driving a single stage centrifugal air compressor via a common rotor shaft to form a self-contained free running unit. Expansion of the exhaust gas through the nozzles results in a high velocity gas stream entering the moving blade assembly. Because of the high rotational speeds perfect dynamic balance is essential if troublesome vibrations are to be avoided. Even with this, the effect of external vibrations being transmitted via the ship's structure to the turbocharger is a further problem to be resolved. This is done by mounting the bearings in resilient housings incorporating laminar spring assemblies to give both axial and radial damping effect. Another aspect of this arrangement is to prevent flutter or chatter at bearing surfaces when stopped so that incidental bearing damage is prevented. Lubrication of the bearings may be by separate or integral oil feed, but whatever arrangement is adopted it must be fully effective at a steady axial tilt of 15° and support a temporary tilt of 22 1/2° as may occur in a heavy seaway. The bearings themselves may be a combination of ball and roller bearings or separate sleeve (journal) type bearings.

The various claims of superiority as to the effectiveness of the different types of bearings centre around the mechanical efficiency of the bearing configuration. The manufacturers of turbochargers equipped with rolling element bearing claim a distinct mechanical efficiency advantage across the whole operating range. On the other hand manufacturers of turbochargers equipped with sleeve type bearings claim comparable efficiency under full-load conditions but accede to lower efficiency at lower engine loads. With high speeds of operation the mechanical efficiency factor does seem to favour rolling element bearings. Against this however is the fact that periodic replacement of ball and roller assemblies is essential if trouble-free service is to maintained – this is due to the fact that rapid and repeated deformation with resultant stressing causes surface metal fatigue of contact surfaces with the result that failure will occur. The effects of vibration, overloading, corrosion or possible abrasive wear, etc, lead to premature failure which emphasises the need for isolation of bearings from external vibrations together with use of correct grade of lubricant and effective filtration, etc. Plain bearings should however have a life equal to that of the blower provided that normal operating conditions are not exceeded.

Referring to Fig. 84. it can be seen that the blower end of the turbocharger consists of a volute casing of light aluminium alloy construction which houses the inducer, impeller and diffuser which are also of light alloy construction. The function of the inducer is to guide the air smoothly into the eye of the impeller where it is collected and flung radially outward at ever increasing velocity due to the centrifugal effect at high rotational speed. At discharge from the impeller it passes to the diffuser where its velocity is reduced in the divergent passages thus converting its kinetic energy into pressure energy. The diffuser also functions to direct air smoothly into the volute casing which continues the deceleration process with further increase in air pressure. From here the air passes to the charge air receiver via the air cooler.

The turbine end of the turbocharger consist of casings which house the nozzle-ring turbine wheel and blading, etc. In older designs casings were water cooled but in turbochargers for modern large slow-speed 2-stroke engines, with relatively low exhaust gas

FIG 84
TURBO-CHARGER

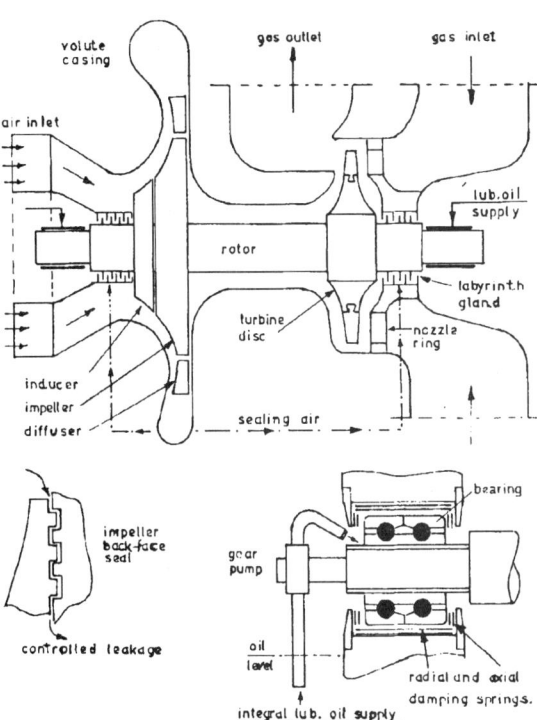

temperatures the casings are uncooled. Uncooled designs retain more heat energy in the exhaust gas in the waste heat boiler so improving the overall plant efficiency. Fig. 85. shows the temperature advantage of uncooled designs.

The components in the high temperature gas stream, that is, the nozzle ring, turbine wheel, blades and rotor shaft are manufactured from heat resisting nickel-chrome alloy steel to withstand continuous operation at temperatures in excess of 450°C. Some degree of cooling may be given by controlled air leak-off past the labyrinth seal, between the back of the impeller and volute casing, which flows along the shaft towards the turbine end.

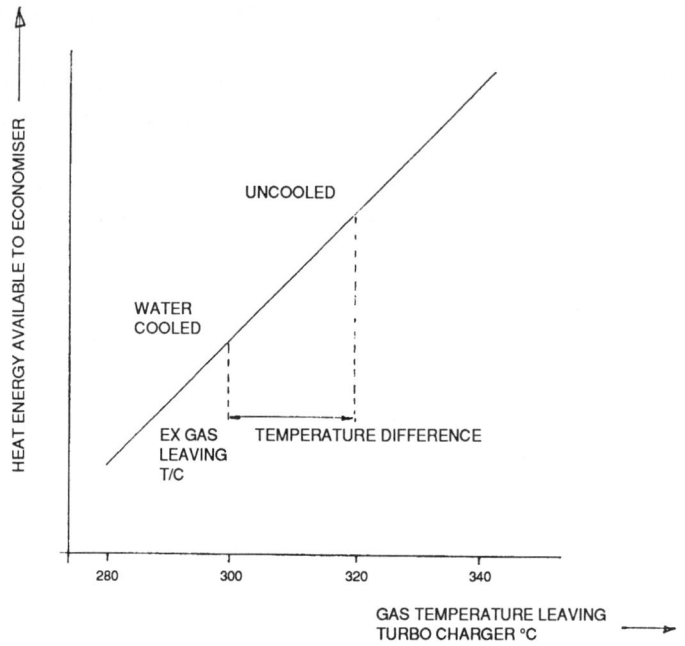

**FIG 85
ADVANTAGE OF UNCOOLED TURBOCHARGER**

Cooling media for cooled exhaust gas casings is generally from the engine jacket water cooling system although in some cases sea-water has been employed. In both cases anti-corrosion plugs are fitted to prevent or inhibit corrosion on the water-side. With water cooled casings experience has shown that under light load conditions when low exhaust temperatures are encountered it is possible that precipitation of corrosive forming products – mainly sulphuric – will occur on the gas side of the casing. This results in serious corrosive attack which is more marked at the outlet casing because of lower temperatures. Methods of prevention such as enamelling and plastic coatings, etc. have been tried to alleviate this problem with varying degrees of success. A particularly effective approach to the problem is the use of air as the cooling media with the result that this particular instance of corrosive attack is virtually eliminated.

Some manufactures utilising sleeve type bearings mount them inboard of the compressor and turbine. This has several advantages:
1. A short, rigid shaft is possible.
2. It allows large volume turbine and compressor inlet casings, free of bearing housings.
3. The main casing, bearing housings and turbo-machinery form one module allowing the rotor to be withdrawn from the turbine casing without disconnecting engine ductwork.

The oil for the bearings is supplied from the main engine lubricating oil system or a separate oil feed as shown in Fig. 86. The oil level in the high level tank should be maintained about 6m

**FIG 86
TURBO-CHARGER LUBRICATION SYSTEM**

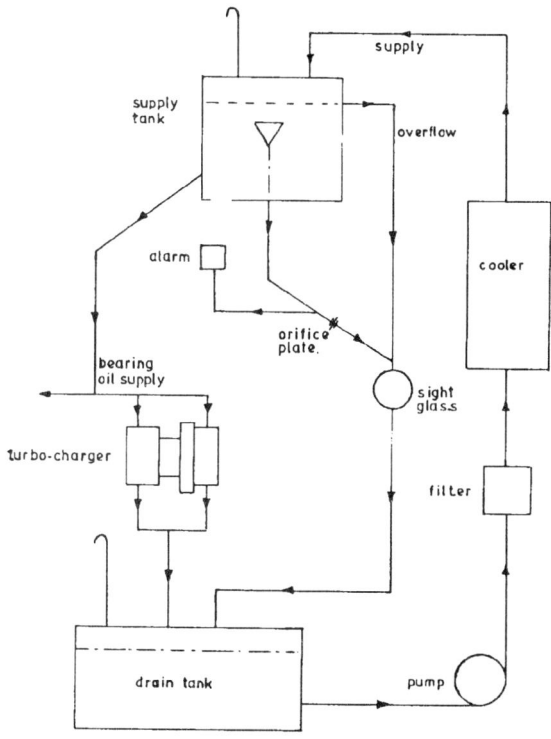

above the turbochargers. This will ensure that the oil pressure reaching the bearings should never fall below a pressure of around 1·6 bar. If level of oil falls below the mouth of the inner drain pipe it is quickly emptied and an alarm condition initiated. After an alarm it takes about ten minutes to empty the high level tank which is sufficient to ensure adequate lubrication of the turbochargers as they run down after the engine is stopped.

As discussed earlier sleeve type bearings suffer the disadvantage of having a lower mechanical efficiency at part load conditions. The effects of this can be minimised by careful design. To reduce friction the bearing length is reduced. A thrust bearing is incorporated into the main bearing but axial thrust is taken by this only at start-up, shut-down and very low loads. The main thrust being taken by sealing air acting on the turbine disc. Fig. 87. shows sealing air from the compressor outlet being fed to the chamber behind the turbine disc. This air flows past the leak-off labyrinth at a rate dependant upon the clearance. As the turbocharger load increases so does the axial thrust. This has the effect of moving the rotating element towards the compressor end which causes the clearance at the leak-off labyrinth to decrease reducing the flow of air. The air pressure acting on the turbine disc increases and imposes an opposing force to the axial force. The makers of turbochargers claim that engines utilising this type of turbocharger can run down to 25% load unassisted by auxiliary fans.

Recent developments have increased the overall efficiency of turbochargers by improving the aerodynamic performance and increases in pressure ratio. One improvement attained is as a result of the general adoption of constant pressure charging for large slow-speed 2-stroke engines. This eliminates the excitation of blade vibration by exhaust gas pulses. Excitation of blade vibration is still possible but with careful attention to the choice of nozzle vane number and natural frequencies of vibration of blades it is possible to dispense with the need for rotor blade damping wire. Not only does this give greater turbine aerodynamic efficiency, but greater resistance to contamination by heavy fuel combustion products.

Radial Flow Turbines
For smaller higher speed diesel applications the use of radial flow turbochargers is common. Fig. 88. The casings are uncooled but

FIG 87
TURBO-CHARGER WITH PLAIN BEARINGS

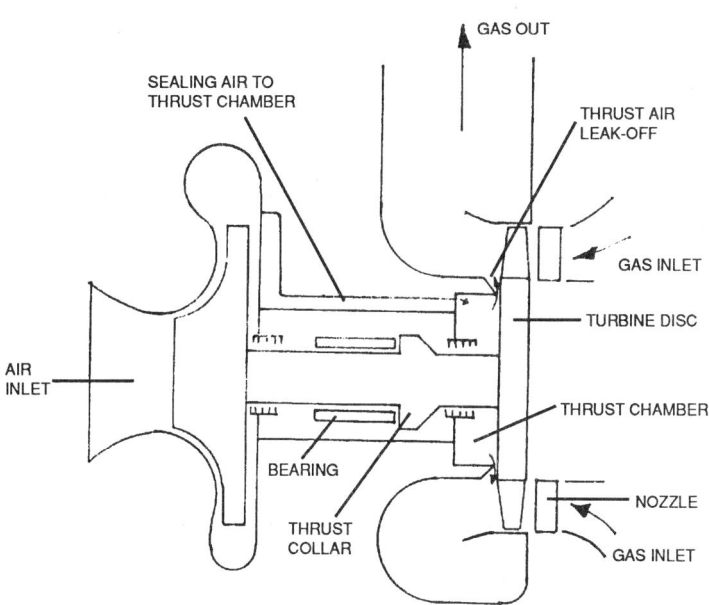

require insulation. Bearings are sleeve type and lubricated form the engine lubricating oil system. The turbine wheel is a one piece casting of a design which gives acceptable efficiencies over the entire operation range. The compressor is also of a one piece design of backswept vane design giving stable operating characteristics. At high air flows the efficiency tends to decrease due to losses at the turbine exit. A comparison between the efficiencies of axial and radial turbines can be seen in Fig. 89.

FIG 88
RADIAL FLOW TURBINE

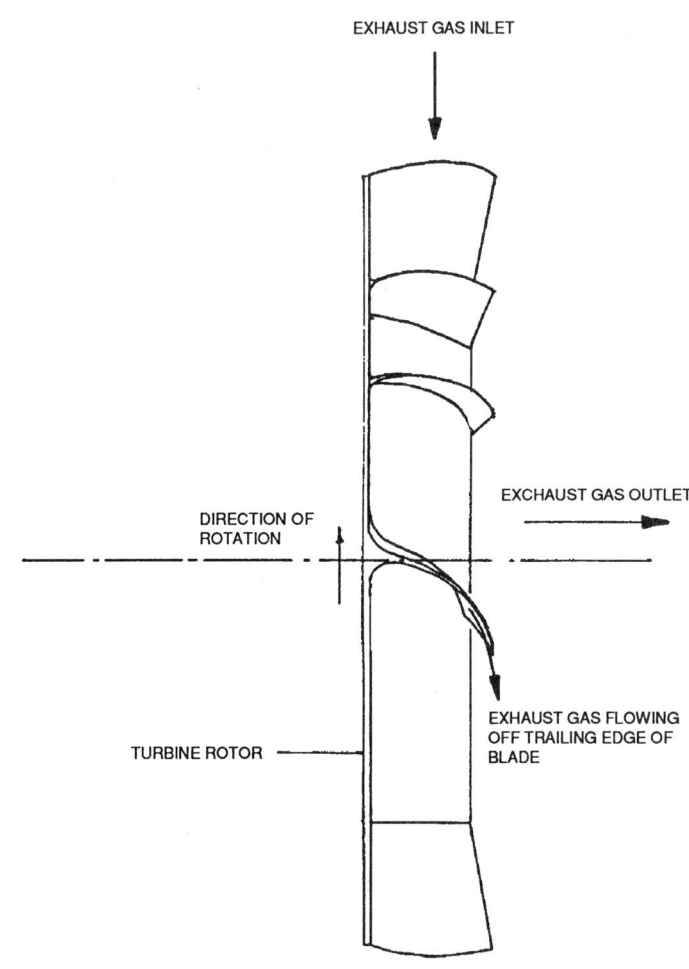

FIG 89
COMPARISON OF RADIAL & AXIAL FLOW T/C EFFICIENCY

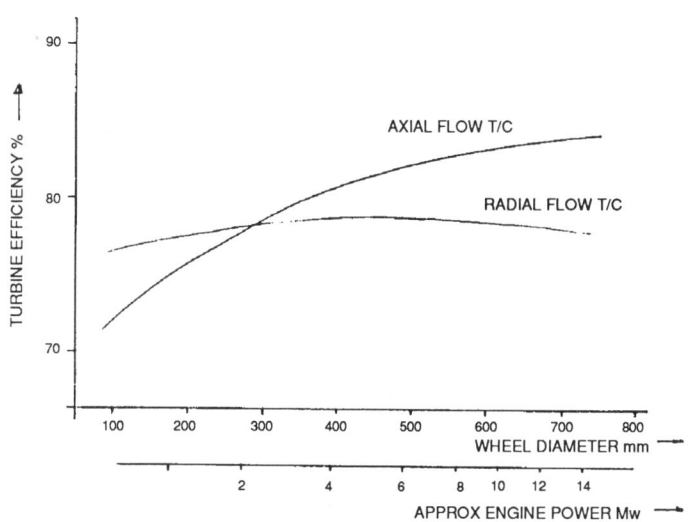

Turbocharger Fouling

Contaminated turbines and compressors have poorer efficiency and lower performance which results in higher exhaust temperatures. In 4-stroke applications the charging pressure can increase due to the constriction of the flow area through the turbine resulting in unacceptable high ignition pressures. To maintain turbocharger efficiency it is important to ensure that all operating parameters are maintained to manufacturers recommendations. If the compressor draws its air from the machinery spaces then steps must be taken to maintain as clean an atmosphere as possible since leaking exhaust gas and/or oil vapour will accelerate the deterioration of efficiency. In some installations the turbo-chargers draw air through ducts from outside the engine room.

TURBOCHARGER CLEANING

Water Washing – Blower Side

On the air side, dry or oily dust mixed with soot and a possibility of salt ingestion from salt laden atmosphere can lead to deposits which are relatively easy to remove with a water jet, usually injected at full load with the engine warm. A fixed quantity of liquid (1 to 2 1/2 litre depending upon blower size) is injected for a period of from 4 to 10 seconds after which an improvement should be noted. If unsuccessful the treatment can be repeated but a minimum of ten minutes should be allowed between wash procedures. Since a layer of a few tenths of a millimetre on impeller and diffuser surfaces can seriously affect blower efficiency the importance of regular water washing becomes obvious. It is essential that the water used for wash purposes comes from a container of fixed capacity – under no circumstances should a connection be made to the fresh water system because of the possibility of uncontrolled amounts of water passing through to the engine.

Water Wash – Turbine Side

This is generally carried out at reduced speed by rigging a portable connection to the domestic fresh water system and injecting water, via a spray orifice before the protective grating at turbine inlet, for a period of 15 to 20 minutes with drains open to discharge excessive moisture which does not evaporate off. Since water washing may not completely remove deposits, and can interact with sulphurous deposits with resultant corrosive attack, chemical cleaning may be used in preference. This effectively removes deposits at the turbine and moreover is still active within the exhaust gases passing to the waste heat system, so that further removal of deposits occurs which maintains heat transfer at optimum condition and keeps back-pressure of exhaust system to well within the limits required for efficient engine operation.

Dry Cleaning – Turbine Side

Instead of water, dry solid bodies in the form of granules are used for cleaning. About 1·5-2 kg of granules is blown by compressed air into the exhaust gas lines before the gas inlet casing or protection grid.

Agents particularly suited to blasting are natural kernel granules, or broken or artificially shaped activated carbon particles with a grain size 1·2 to 2·0 mm.

The blasting agents have a mechanical cleaning effect, but it is not possible to remove fairly thick deposits with the comparatively small quantity used. For this reason this method must be adopted more frequently than for cleaning with water. Dry cleaning is carried out every 24 to 50 hours. The main advantage of this type of turbine cleaning is that it can be carried out at full or only slightly reduced load. The cleaning equipment configuration is shown in Fig. 90.

Turbocharger manufacturers recommend that heavily contaminated machines, which have not been cleaned regularly from the very beginning or after overhaul, should not be cleaned by water-washing or granulate injection. This is because the dangers of incomplete removal of deposits may cause rotor imbalance.

Surging

Surging is a phenomenon that affects centrifugal compressors when the mass flow rate of air falls below a sustainable level for a given pressure ratio. Consider the system in Fig. 91. where a constant speed compressor supplies air though a duct. The outlet of the duct is regulated by a damper. With the damper fully open the pressure ratio across the compressor will be at its lowest value with the largest mass flow rate of air. As the damper is closed the resistance increases as does the pressure ratio but the mass flow of air decreases. If the damper is closed further a point will be reached where, because of the resistance, there will be such a low mass flow rate and high pressure ratio across the compressor that flow breaks down altogether. When this occurs the pressure downstream of the compressor is simply relieved to atmosphere, backwards, through the compressor. This is known as surging and is accompanied by loud sounds of "howling and banging". The

FIG 90
DRY TURBO-CHARGER CLEANING EQUIPMENT

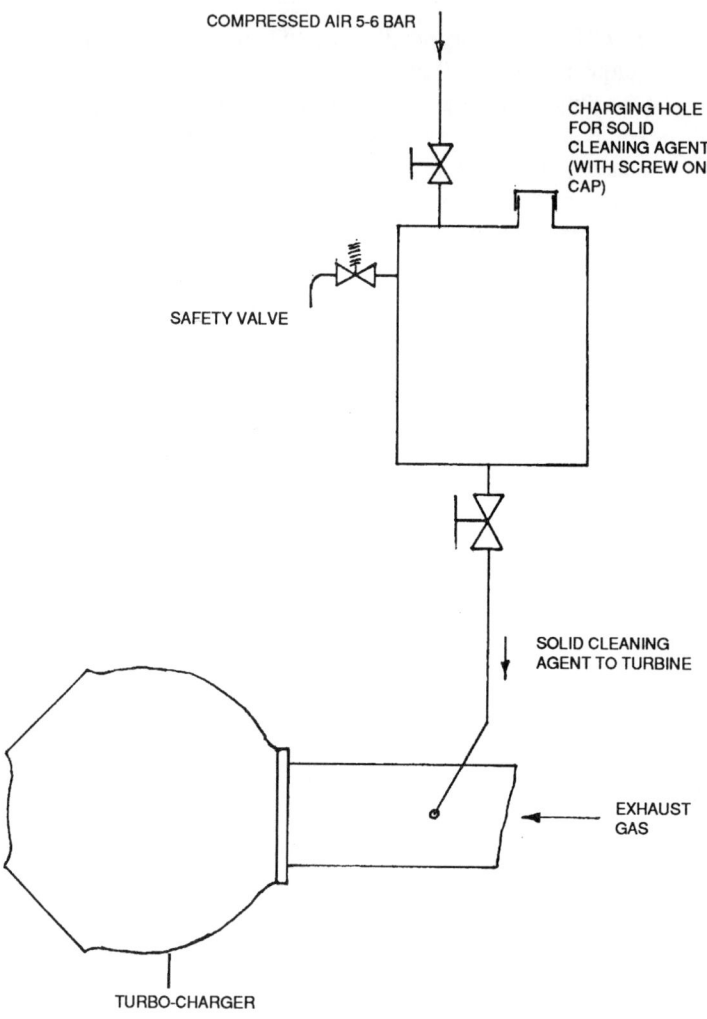

FIG 91
SURGING OF TURBO-CHARGER COMPRESSOR

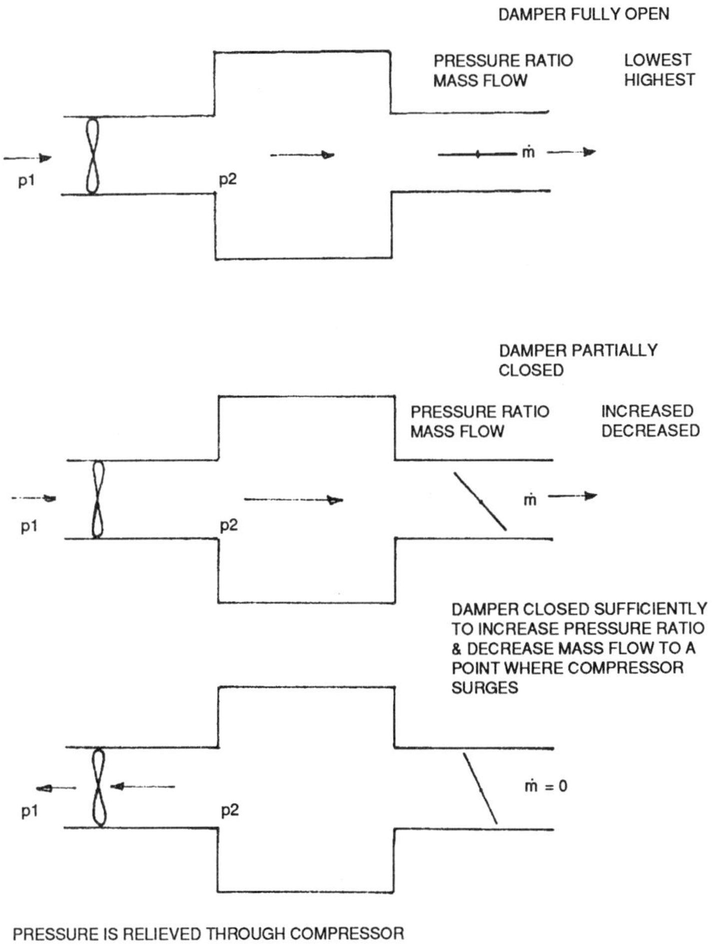
DAMPER FULLY OPEN
PRESSURE RATIO LOWEST
MASS FLOW HIGHEST
p1
p2
ṁ
DAMPER PARTIALLY CLOSED
PRESSURE RATIO INCREASED
MASS FLOW DECREASED
p1
p2
ṁ
DAMPER CLOSED SUFFICIENTLY TO INCREASE PRESSURE RATIO & DECREASE MASS FLOW TO A POINT WHERE COMPRESSOR SURGES
p1
p2
ṁ = 0
PRESSURE IS RELIEVED THROUGH COMPRESSOR

events leading to the surging can be followed on a graph of pressure ratio against mass flow. This graph is known as a compressor map. Fig. 92b.

Surging may occur in heavy weather when the propeller comes out of the water and the governor shuts the fuel off almost instantaneously.

To obtain efficient and stable operation of the charging system it is essential that the combined characteristic of the engine and blower are carefully matched. The engine operating line, as indicated on Fig. 92a., is mainly a function of these characteristics and taking into account the fact that blower efficiency decreases as the distance between surge and operating lines increases, the matching of blower to engine becomes a compromise between acceptable blower efficiency and a reasonable safety margin against surge. An accepted practice is to provide a safety margin of around 15 to 20% to allow for deterioration of service conditions such as fouling and contamination of turbochargers and increasing resistance of ship's hull, etc. Apart from fouling of turbocharger other contributory factors to surging are contamination of exhaust and scavenge ducting, ports and filters. Since faulty fuel injection leads to poor combustion and greater release of contaminants the need to maintain fuel injection equipment at optimum conditions is essential. Other related causes are variation in gas supply to turbochargers due to unbalanced output from cylinder units and

**FIG 92A
PROVIDING A SAFETY MARGIN**

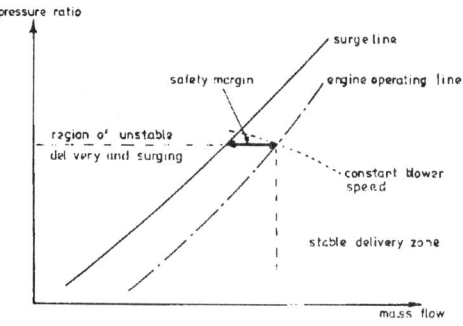

**FIG 92B
COMPRESSOR MAP**

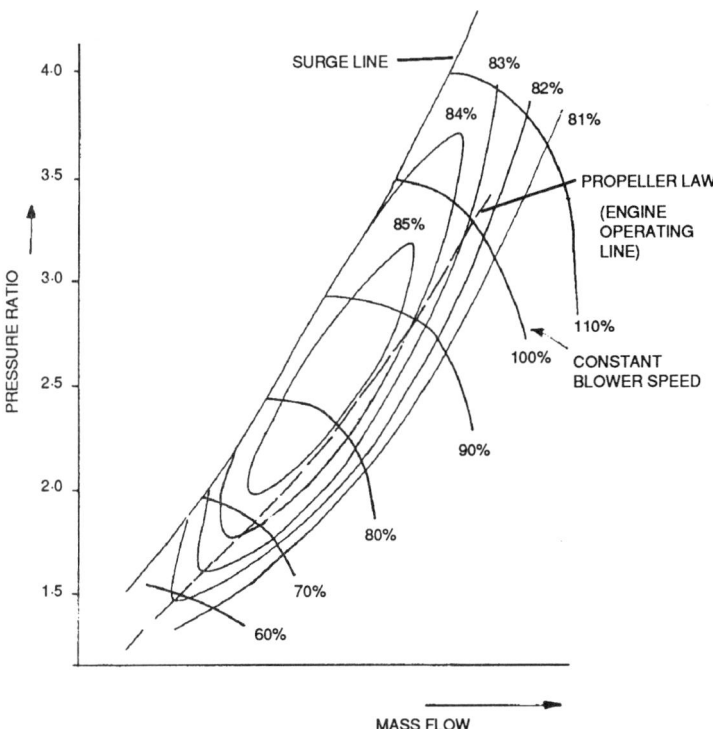

mechanical damage to turbine blading, nozzles or bearings, etc.

During normal service the build-up of contaminants at the turbocharger can be attributed to deposition of air-borne contaminants at the compressor which in general are easily removed by waterwashing on a regular basis. At the turbine however, more active contaminants resulting from vanadium and sodium in the fuel together with the products of incomplete combustion deposit at a higher rate which increases with rising temperature. A further problem arises with the use of alkaline cylinder lubricants with the formation of calcium sulphate deposits originating form the alkaline additives in the lubricant. Again water washing on a regular basis is beneficial in removing and controlling deposits but particular care needs to be taken to ensure complete drying out after the washing sequence since any remaining moisture will interact with sulphurous compounds in exhaust gas stream with damaging corrosive effect.

Turbocharger Breakdown

For correct procedure, depending upon engine type, reference should be made directly to the engine builders and/or turbocharger manufacturers recommended practice. As a general rule however, in the event of damage to the turbochargers, the engine should be immediately stopped in order that the damage is limited and does not become progressive. Under conditions where the engine cannot be stopped, without endangering the ship, engine speed should be reduced to a point where turbocharger revolutions have fallen to a level at which the vibration usually associated with a malfunction is no longer perceptible.

If the engine can be stopped but lack of time does not permit *in situ* repair or possible replacement of defective charger it is essential that the rotor of the damaged unit is locked and completely immobilised. If exhaust gas still flows through the affected unit once the engine is restarted, the coolant flow through the turbine casing needs to be maintained but due to the lack of sealing air at shaft labyrinth glands the lubricating oil supply to the bearings will need to be cut off – with integral pumps mounted on the rotor shaft, the act of locking the shaft ensures this – otherwise contamination of lubricant together with increase in fouling will occur. For rotor and blade cooling a restricted air supply is required and can be achieved by closing a damper or flap valve in the air delivery line from the charger, to a position which gives limited flow from scavenge receiver back to the damaged blower. Alternatively a bank flange incorporating an orifice of fixed diameter can be fitted at the outlet flange of the blower.

With only a single blower out of a number inoperative the power developed by the engine will obviously depend upon charge air pressure attainable. At the same time a careful watch must be kept upon exhaust condition and temperature to ensure efficient engine operation with good fuel combustion. In the event of all turbochargers becoming defective it is possible to remove blank covers from the scavenge air receiver so that natural aspiration supplemented by underpiston effect, etc. or parallel auxiliary blower operation is possible – if this method of emergency operation is carried out protective gratings must be fitted in place of blind covers at the scavenge air receiver. In all cases when running at reduced power special care must be taken to ensure any out of balance due to variation in output from affected units does not bring about any undue engine vibration.

CHAPTER 5

STARTING AND REVERSING

Starting Air Overlap
Some overlap of the timing of starting air valves must be provided so that as one cylinder valve is closing another one is opening. This is essential so as to ensure that there is no angular position of the engine crankshaft with insufficient air turning moment to give a positive start. The usual minimum amount of overlap provided in practice is 15°. Starting air is admitted on the working stroke and the period of opening is governed by practical considerations with three main factors to consider:

1. The Firing Interval of the engine.

$$\text{Firing Interval} = \frac{\text{Number of degrees in engine cycle}}{\text{Numbers of cylinders}}$$

e.g. with a four cylinder 2-stroke engine the firing interval is 90°, *i.e.* 360/4 and if each cylinder valve covered 90° of the cycle then the engine would not start if it had come to rest in the critical position with one valve fractionally off closure and another valve just about to start opening.
2. The valve must close before the exhaust commences. It is rather pointless blowing high pressure air straight to exhaust and it could be dangerous.
3. The cylinder starting air valve should open after firing dead centre to give a positive turning moment in the correct direction. In fact some valves are arranged to start to open as much as 10° before the dead centre because the engine is past this position before the valve is effectively well open, and in fact any reverse turning effect is negligible as the turning moment exerted on a crank very near dead centre is small indeed.
   Consider Fig. 93(a). for a 4-stroke engine. With the timings as shown the air starting valve opens 15° after dead centre and closes 10° before exhaust begins. The air start period is then 125°. The

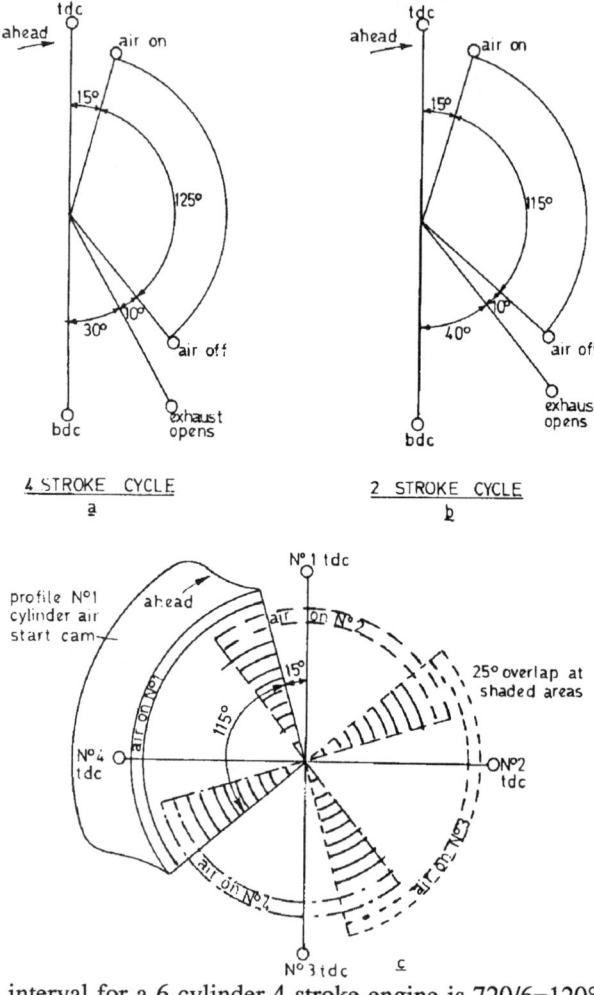

**FIG 93
AIR START CAM AND CRANK TIMING DIAGRAMS**

firing interval for a 6-cylinder 4-stroke engine is 720/6=120°. The period of overlap is 5° which is insufficient. Although this example could easily be modified so as to give sufficient (say 15°) overlap by reducing the 15° after dead centre and the 10° before exhaust opening, it can become very difficult to arrange with very early exhaust opening on turbocharged engines. A 7-cylinder 4-stroke engine is much easier to arrange.

Consider Fig. 93(b). for a 2-stroke engine:
This has an air start period of 115°. Firing interval for a 3-cylinder 2-stroke engine = 360/3 = 120°. This means no overlap. Modification can arrange to give satisfactory starting with this example but for modern turbocharged 2-stroke engines having exhaust opening as early as 75° before bottom (outer) dead centre it becomes virtually impossible. A 4-cylinder 2-stroke engine is much easier to arrange and would be adopted. Consider Fig. 93(c). which is a cam diagram for a 2-stroke engine with 4 cylinders. The air open period is 15° after dead centre to 130° after dead centre, *i.e.* a period of 115°. This gives 25° of overlap (115 - 360/4) which is most satisfactory. Take care to note the direction of rotation and this is a cam diagram so that for example, No. 1 crank is 15° after dead centre when the cam would arrange to directly or indirectly open the air start valve. The firing sequence for this engine is 1 4 3 2. This is very much related to engine balancing and no hard and fast rules can be laid down about crank firing sequences as each case must be treated on its merits.

It may be useful to note that for 6-cylinder, 2-stroke engines a very common firing sequence is 1 5 3 6 2 4 and similarly for 7- and 8-cylinders 1 7 2 5 4 3 6 and 1 6 4 2 8 3 5 7 respectively are often used.

The cam on No. 1 cylinder is shown for illustration as it would probably be for operating say cam operated valves, obviously the other profiles could be shown for the remaining three cylinders in a similar way. The air period for Nos. 1, 4, 3 and 2 cylinders are shown respectively in full, chain dotted, short dotted and long dotted lines and the overlap is shown shaded.

Starting Air Valves
Each cylinder is fitted with a starting air valve which is operated pneumatically by one of the starting air distributor control valves. Fig. 94. These are arranged radially around the starting air distributor cam as shown in Fig. 95(a). At the correct engine position an air signal from the control valve is directed to the cylinder starting air valve upper chamber and acts on the piston to

FIG 94
STARTING AIR VALVE

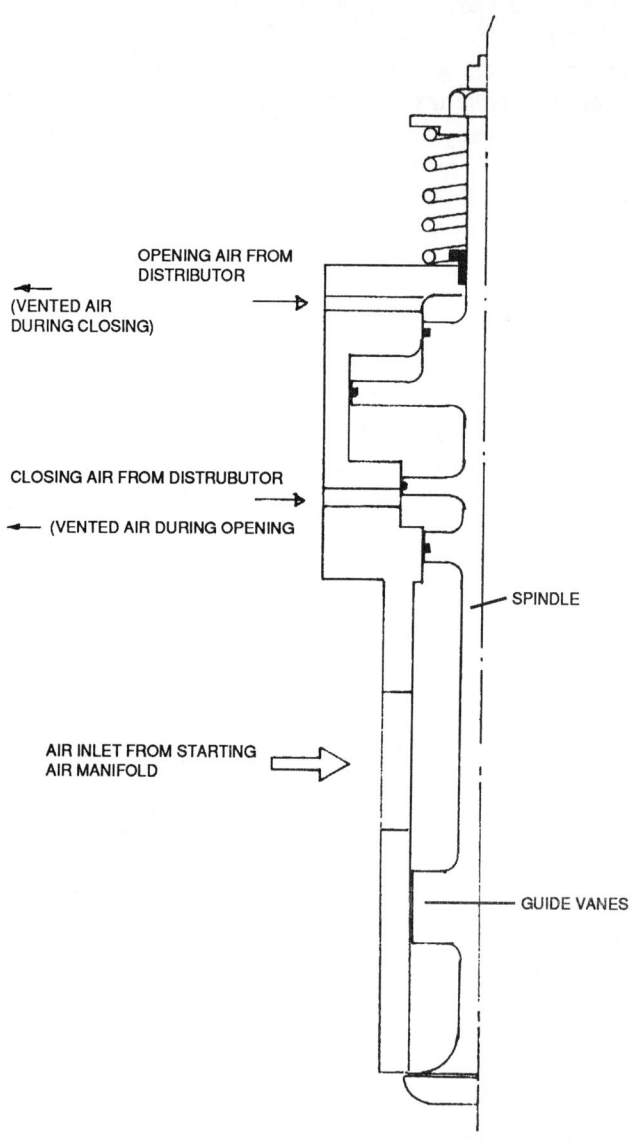

FIG 95A
STARTING CONTROL VALVE ARRANGEMENT FOR EIGHT CYLINDER ENGINE (SULZER TYPE)

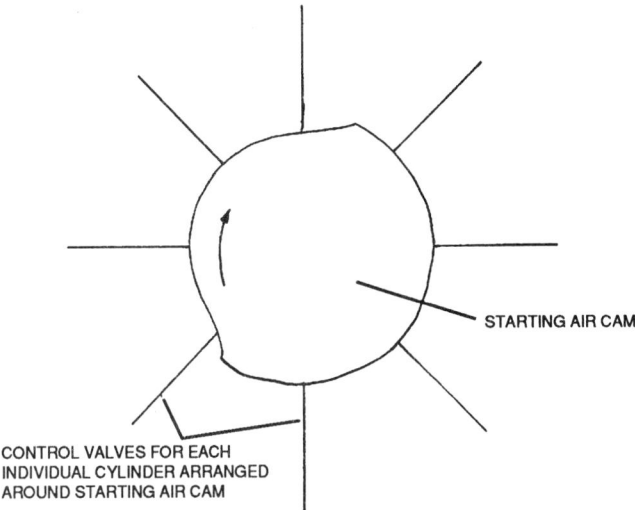

open the valve. As this is happening the air from the lower chamber is vented to atmosphere through the control valve. At the end of the starting air admission period control air is redirected to the lower chamber to close the valve while the upper chamber is vented to atmosphere through the control valve. The valve opens and closes quickly with air cushioning at the end of the closing motion to reduce shock on the valve seat. If the pressure in the cylinder is substantially higher than the starting air pressure, the valve will not open. This prevents hot gases entering the starting air manifold.

During engine operation the air inlet to the starting valve should be regularly checked. A hot inlet would indicate a leaking starting air valve allowing hot combustion gases to enter the air manifold which may lead to an explosion if starting air is admitted.

Starting Air Distributor
There are many designs of air distributor all with the same basic principles, *i.e.* to admit air to the pistons of cylinder relay valves in the correct sequence for engine starting. Valves not being supplied with air would be vented to the atmosphere via the distributor. Some overlap of timing would obviously be required.

One type of starting air distributor is shown in Fig. 95. This is based on the Sulzer design in which each cylinder has its own starting control valve. The starting control valves are arranged radially around the starting air distributor cam, which is driven via a vertical shaft from the camshaft. When the engine starting lever is operated air is admitted to the distributor forcing all control valves, against the return spring, [omitted in the diagram] onto the cam. The control valve of a cylinder which is in the correct position for starting, will be pushed into the depression in the cam and assume the position shown in Fig. 95(b). In this position air from the starting system will be directed to the upper part of the cylinder starting air valve causing the valve to open. At the same time air from the lower chamber of the cylinder starting air valve will be vented to atmosphere. At the end of the cylinder starting air period the distributor cam moves the control valve to the position shown in Fig. 95(c). In this position air from the starting air system is directed to the lower chamber of starting air valve causing the valve to close. Air from the upper chamber is vented through the control valve to atmosphere. The starting control valves are held off the distributor cam by springs when starting air is shut off the engine.

General Reversing Details

Most reversible engines are of the direct coupled 2-stroke type. The general trend in 4-stroke practice is to utilise an unidirectional engine coupled, via a reduction gearbox, to a controllable pitch propeller. The need for reversing mechanisms for 4-stroke engines is, therefore, reducing. For this reason the 2-stroke reversing mechanism principle will be considered in greatest detail.

2-Stroke Reversing Gear

It is usually necessary to reposition the fuel cams on the camshaft, with jerk pumps, so that reversing can utilise one cam. This avoids the complication of moving the camshaft axially. This means that it is necessary to provide a lost motion clutch on the camshaft and the need for such a clutch will first be described.

Referring to Fig. 96. the lost motion cam diagram:

Consider the engine position to be dead centre ahead with the cam peak centre line to be 55° after this position, anti-clockwise ahead rotation, for correct injection timing ahead. If now the

FIGS 95B, C
STARTING AIR DISTRIBUTOR

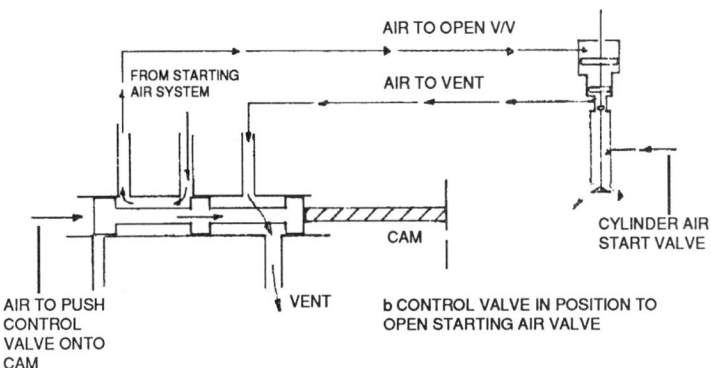

b CONTROL VALVE IN POSITION TO OPEN STARTING AIR VALVE

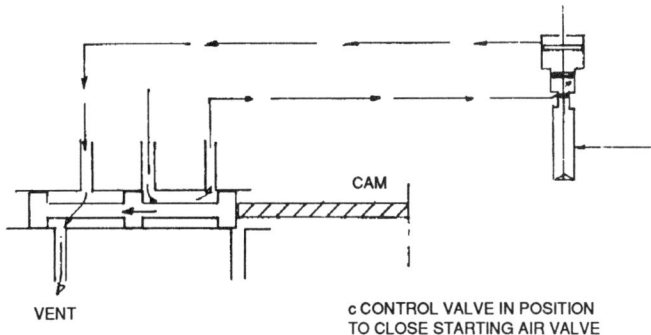

c CONTROL VALVE IN POSITION TO CLOSE STARTING AIR VALVE

engine is to run astern (clockwise) the cam is $55 + 55 = 110°$ out of phase. Either the cam itself must be moved by 110° or while the engine rotates 360° the cam must only rotate 250° (110° of lost motion). Note the symmetrical cam 75° each side of the cam peak centre line made up of 35° rising flank and 40° of dwell.

FIG 96
LOST MOTION CAM DIAGRAM

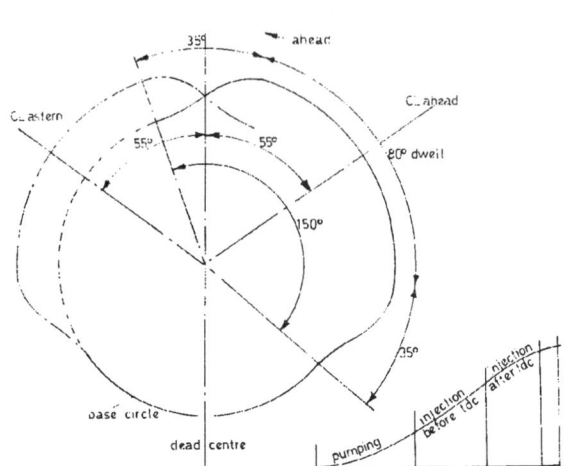

The flank of the cam is shown on an enlarged scale in Fig. 96. It will be noted that the 35° of cam flank is utilised for building up pressure by the pumping action of the rising fuel pump plunger (14°) for delivery at injection 10° before firing dead centre to 8° after firing dead centre, and 3° surplus rise of flank for later surplus spill variation. It is obvious that the lost motion is required with jerk pumps, cam driven, in which a period of pumping is necessary before injection starts. The following points are worth specific mention:

1. Pielstick and MAN 2-stroke engines have ahead and astern cams on the same shaft. To change from one set of cams to the other, the camshaft is moved axially and so no lost motion is required.

2. The dwell period is not normally necessary from the fuel injection aspect alone, *i.e.* about 30° lost motion would be adequate and is provided as such on British Polar and older Sulzer engines.

3. Dwell, in which the fuel plunger is held before return is often provided to give a delay interval. For example with older B & W. engines about 80° dwell gives a rotation (total) of the camshaft of

about one third of a revolution which allows an axial travel with a screw nut arrangement of reasonable size and pitch to change over for reverse running.

4. Older loop scavenged Sulzer engines have about 98° lost motion as the distributor repositioning for astern is from the same drive shaft as the fuel pumps, but via a vertical direct drive shaft.

Refer now to Fig. 97. the Lost Motion Clutch.

This design which is based on older Sulzer engine practice has a lost motion on the fuel pump camshaft of about 30°. When reversal is required oil pressure and drain connections are reversed. Oil flowing laterally along the housing moves the centre section to the new position, *i.e.* anticlockwise as shown on the sketch in Fig. 97. The oil pressure is maintained on the clutch during running so that the mating clutch faces are kept firmly in contact with no chatter.

There are a number of variations on this design but the principle of operation is similar although not all types rotate the clutch to its new position before starting and merely allow the camshaft to "catch on" with the crankshaft rotation when lost motion is completed.

FIG 97
LOST MOTION CLUTCH (EARLY SULZER)

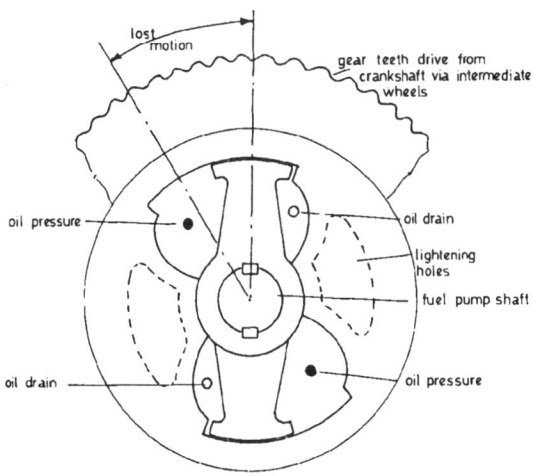

PRACTICAL SYSTEMS

Having described the basic principles of starting and reversing the actions are now combined to give a selection of systems as used on the various engine types.

Starting Air System (B & W.)

Consider first the air off position. Air from the storage bottle passes to the automatic valve which however remains shut as air passes through the pilot valve (1) to the top of the automatic valve piston. All cylinder valves and distributor valves are venting to atmosphere via the automatic valve. If now the lever is moved to the position shown in Fig. 98. then the air pressure on top of the automatic valve is vented through the pilot valve (1) by the linkage shown. This causes the automatic valve to open as the up force on

**FIG 98
STARTING AIR SYSTEM (B & W)**

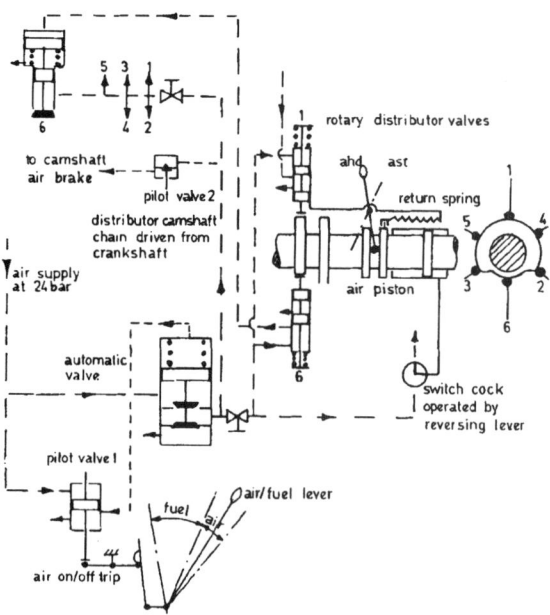

the larger piston is greater than the down force on the smaller valve with the spring force. The lower vent connection is closed and air flows to all cylinder and distributor valves.

The cylinder valves are of the air piston relay type described earlier and in spite of main air pressure on them will be closed except for one valve (or possibly two). This distributor has the piston pilot valves mounted around the circumference of a negative cam. Only one distributor pilot valve can be pushed into the negative cam slot, *i.e.* No. 6. and hence air flows through the No. 6. distributor pilot valve only to the upper part of the piston for the No. 6. cylinder air starting valve, which will open. All other starting valves are shut and venting to atmosphere. The position shown for illustration is air on to No. 6. cylinder of a 6-cylinder engine running ahead. When the lever is moved forward on to fuel the whole system is again vented to atmosphere through the automatic valve.

For astern running the reversing lever is moved over which allows air to pass, via a switch cock, to push the light distributor shaft along by means of an air piston (alternatives are scroll, direct linkage, etc.), so putting the astern distributor cam into line with the distributor pilot valves. Distributor pilot valves are kept out by springs during this operation. The air-fuel lever is then operated as previously described for the engine to run astern. Air start timing for a 2-stroke engine, upon which the above system is typical, is 5° before firing dead centre to 108° after firing dead centre (122° after for astern). B & W. engines also employ a revolving plug type of distributor on some engine designs. Again some types of these engines utilise an air brake on the main camshaft so that air pistons pressure against the pilot valve (2), operated from the reversing lever, while the lost motion is being travelled by the engine. The main camshaft is therefore kept stationary and just before the lost motion is complete the air pressure is released to atmosphere so releasing the brake.

Starting Air System (Sulzer RND).
Refer to Fig. 99.

Air from the starting receiver at 30 bar maximum flows to the pre-starting valve (via the open turning gear blocking valve shown), and directly to the automatic valve. At the automatic valve

FIG 99
STARTING AIR SYSTEM (SULZER RND)

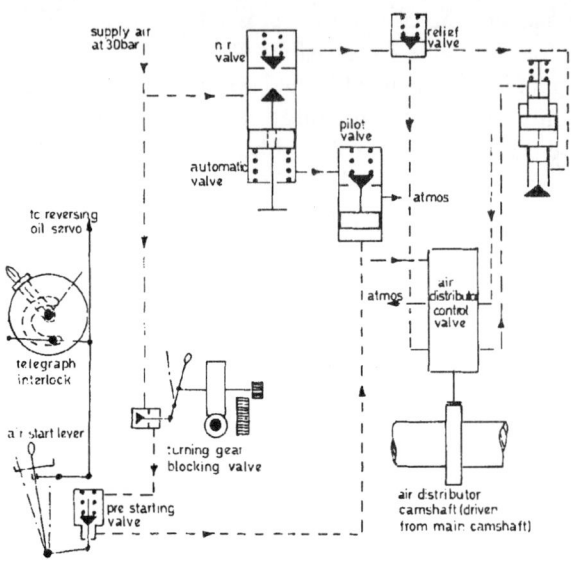

air passes through the small drilled passage to the back of the piston and this together with the spring keeps this valve shut as the pilot valve is shut with air pressure on top and atmospheric vent below.

If the air starting lever is operated with control interlocks free, the opening of the pre-starting valve allows air to lift the pilot valve, vent the bottom of the automatic valve and cause it to open as shown. This allows air to pass to the cylinder valve via non-return and relief valves and also to the distributor. The distributor will allow air to pass to the appropriate cylinder valve causing it to open due to air pressure on the piston top. In this design when the piston top of the cylinder valve is connected to the atmosphere for venting, the bottom of the valve is connected to air pressure, this ensures a rapid closing action. The distributor of this engine is very similar in principle to that shown for the B & W. engine previously except that a positive cam is used by Sulzer.

A mechanical interlock is provided as a blocking device from the telegraph as shown. There is also a connection to the reversing oil servo and an interlock connection from the reversing system to the air start lever via a blocking valve. These are described for the next sketch. Fig. 100.

Hydraulic Control System (Sulzer RND)

Consider a reversing action from ahead to astern.

Oil pressure from left of reversing valves to right of the clutch and under relay valve A. and air block valve.

The telegraph reply lever on the engine telegraph is first moved to stop and the fuel lever moved back to about notch $3_{1/2}$, the starting lever is mechanically blocked by the linkage shown Fig. 99.

The telegraph linkage to the reversing valve moves this valve and releases oil pressure from the lost motion clutch. This drop in pressure causes relay valve A. to move down by spring action which relieves pressure on the block piston (fuel) so cutting off fuel injection. The pressure on the block valve (cam) is also relieved which serves to also lock the starting lever.

Consider now the situation as shown on the sketch of Fig. 100.

When engine speed reduces the telegraph lever can be moved to astern. This allows pressure oil to flow from the right through the reversing valve, as shown on the sketch, to the left of the lost motion clutch to re-position them for astern.

When the servo has almost reached the end of its travel, pressure oil admitted to the block valve (air) releases the lock on the air start lever. (The mechanical lock on the air lever with the telegraph had been released when the telegraph lever was moved to the astern running position.) Pressure oil also acts on relay valve A. admitting oil to block piston (fuel) so allowing the fuel control linkages to the fuel pumps to assume a position corresponding to the load setting of the fuel lever.

If the pressure trips act in the event of low oil pressure (supply and bearings) or low water pressure (jacket or piston) then a trip piston moves up under preset spring pressure so connecting the oil pressure connection to drain. This pressure drop causes the block piston (fuel) to rise up under its spring force and shut off fuel injection.

Connections 1 and 3 from the running direction safety interlock to the reversing valve only allow fuel to the engine if the rotation agrees with the telegraph position. If not, the block piston (fuel) is

FIG 100
HYDRAULIC CONTROL SYSTEM (SULZER RD)

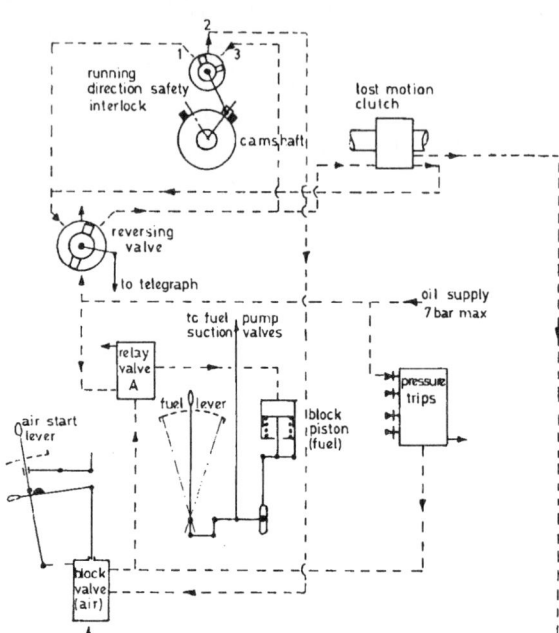

relieved of pressure via block valve (air) and relay valve A.

Movement of the air starting lever can now be carried out as both locks have been cleared and subject to no trip action and satisfactory correspondence between rotation direction and telegraph reply lever indication fuel can be admitted following the full sequence of air starting as described previously, and illustrated in Fig. 99. It is obvious that this system has a large amount of auto-control and is easily adjusted for bridge control.

Control Gear Interlocks

There are many types of safety interlocks on modern IC engine manoeuvring systems. The previous few pages have picked out a number relating to the Sulzer RND engine and these will be adequate to cover most engine type designs as principles are all very similar.

Consider the interlock systems illustrated in Fig. 100. and 101. The telegraph and turning gear interlocks are straight mechanical linkages. In the former case rotation of the telegraph lever from stop position causes the pin to travel in the scroll and unlock the air start lever as well as re-position the reversing valve. The turning gear blocking valve can be seen to close when the pinion is placed in line with the toothed turning gear wheel of the engine. The interlock exerted on the block piston (fuel) is also a fairly simple principle working on the relay valve A. from the pressure trips and is as described previously.

Similarly the block valve (air) operates mechanically via the lever lock on air start lever and horizontal operating lever which rises to unlock under the oil pressure acting through the Servo on block valve (air) after the clutch reversals have taken place. (The pressure trips are merely spring loaded pistons moving against low oil or water pressure to relieve control oil pressure just like conventional relief valves.) It is perhaps appropriate here to describe one trip in detail and the direction safety lock will now be

FIG 101
SAFETY LOCK FOR CORRECT ROTATION (SULZER RND)

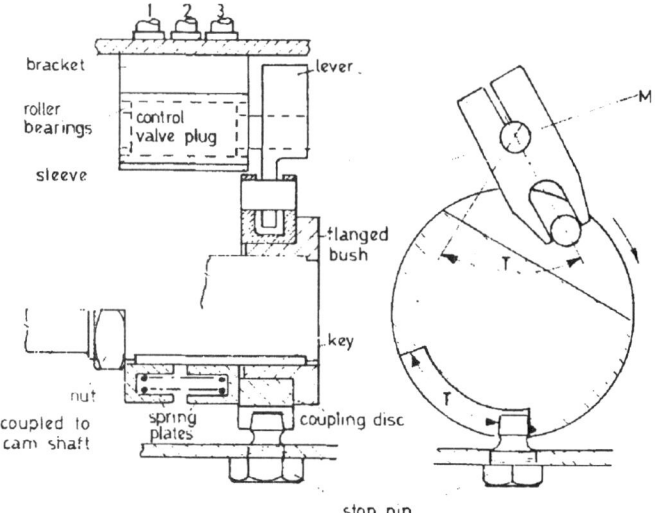

considered briefly. The function is to withhold fuel supply during manoeuvring if the running direction of the engine is not coincident with the setting of the engine telegraph lever. Refer now to Fig. 101.

At the camshaft forward end the shaft is coupled to the cam shaft and carries round with it, due to the key, a flanged bush and spring plates which cause an adjustable friction pressure axially due to the springs and nut. This pressure acts on the coupling disc which rotates through an angular travel T until the stop pin prevents further rotation. This causes angular rotation of a fork lever and the re-positioning of a control valve plug in a new position within the sleeve. Oil pressure from the reversing valve can only pass to the block valve (air) and unlock the air start lever and the fuel control if the rotation of the direction interlock is correct. If the stop pin were to break the fork lever would swing to position M and the fuel supply would be blocked.

Modern Reversing Systems

In the previous sections reversal was carried out by utilising lost motion or by moving the camshaft axially to utilise a different set of cams. It can be seen that these methods involve added complication in the running gear and control systems of engines. In order to eliminate undue complications and improve engine response B & W. have designed a much simpler reversing system, eliminating the need to change the relative positions of exhaust cam and crankshaft.

In the latest L-MC designs, although the exhaust valve opening is not symmetrical about bottom centre (see Fig 13 d), the engine is able to operate in both directions without the necessity of changing exhaust valve position. However, astern operation is somewhat impaired because late closing of the exhaust valve allows a loss of combustion air to exhaust.

In order to change the fuel pump timing for astern running B & W. utilise a movable fuel pump guide roller. Fig. 102(a). shows the guide roller in the ahead position. To change to the astern position a pneumatic cylinder, controlled from the engine starting and reversing system, moves the guide roller to the position shown in Fig. 102(b). This has the effect of correctly positioning the fuel pump for astern running.

FIG 102
REVERSING MECHANISM OF MODERN B&W ENGINES

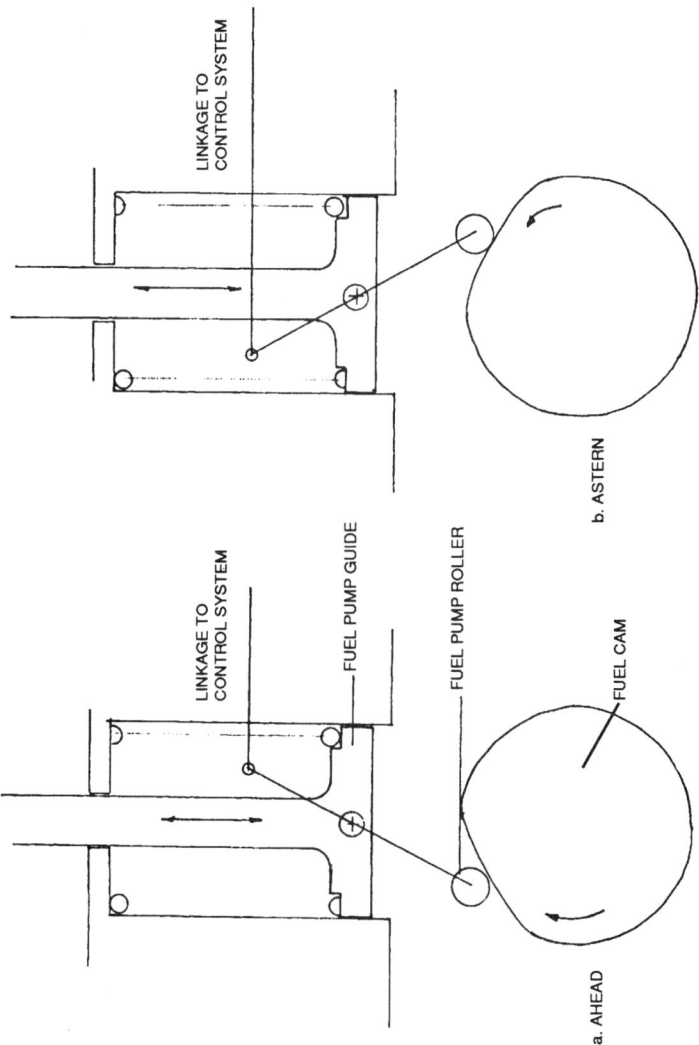

CHAPTER 6

CONTROL

The study and application of instrument and control devices has developed from the beginning of engineering itself. It is however in the last few years that this branch of engineering has assumed greater importance. Automatic Control in a simple sense has always been utilised, *e.g.* cylinder relief valves, speed governors, overspeed trips, etc. It is intended in this chapter to examine the control of the modern diesel engine and its associated equipment and to apply control terminology, with explanations, where required.

GOVERNING OF MARINE DIESEL ENGINES

A clear distinction is necessary between the function of a governor and an overspeed trip. For engine protection governors should not be the only line of defence. While governors control the engine speed between close limits, separate independent overspeed protection is necessary to shut down the engine in the event of the instantaneous shedding of load or governor malfunction.

In the past diesels driving electrical generators invariably utilised plain flyweight governors of the Hartnell type (Fig. 103.) a change in speed resulted in variation of the position of the flyweights and alteration of fuel supply. Larger, slow running direct drive diesel engines were not generally fitted with such a governor but they invariably were fitted with an overspeed trip, usually of the Aspinal inertia type. This trip was arranged to allow full energy supply under normal operating conditions but in the event of revolutions rising about 5% above normal the energy was totally shut off until revolutions dropped to normal again. At about 15% above normal revolutions the trip would stay locked, with energy shut off, and this would continue until re-set by hand.

A plain flyweight governor, which has to perform two separate functions (1) to act as a speed measuring device, (2) to supply the necessary power to move the fuel controlled system. Fig. 104. shows in block diagram form the arrangement. A closed loop

FIG 103
MECHANICAL GOVERNOR

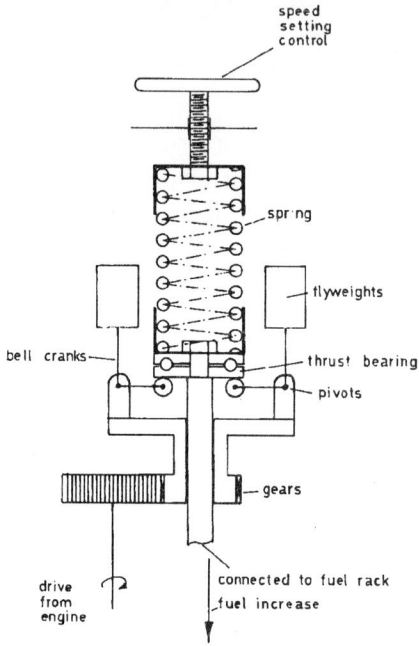

FIG 104
CLOSED LOOP CONTROL

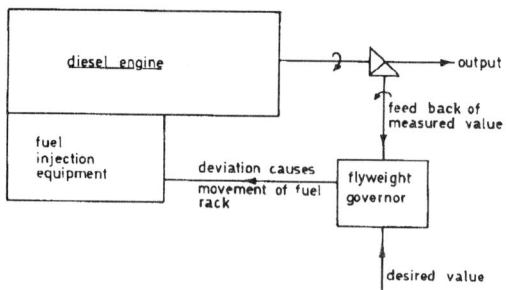

control system is one in which the control action is dependent on the output. The measured value of the output, in this case the engine speed, is fed back to the controller which compares this value with the desired value of speed. If there is any deviation between the values, measured and desired, the controller produces an output which is a function of the deviation. In this case the controller output would be proportional to the deviation, *i.e.* proportional control.

In control terminology deviation is sometimes called error, since it is the difference between measured and desired values, and desired value is sometimes called set value. Proportional control suffers from offset. In the example, if a speed change occurs the flyweights take up a new equilibrium position and the fuel supply will be altered to suit the new conditions. However, the diesel is now running at a slightly different speed to before. If the original speed was the desired value then the new speed is offset from the desired value.

In governor parlance the term speed droop, or just droop, is used to define the change in speed between no load and full load conditions. If speed droop did not exist then there would be one speed only for any position of the governor flyweights and this in turn means any fuel supply rate. In this case the diesel would hunt.

This is an isochronous condition, an engine fitted with an isochronous governor will hunt. However, the term isochronous has taken a new meaning as we will find later.

FIG 105
SPEED DROOP

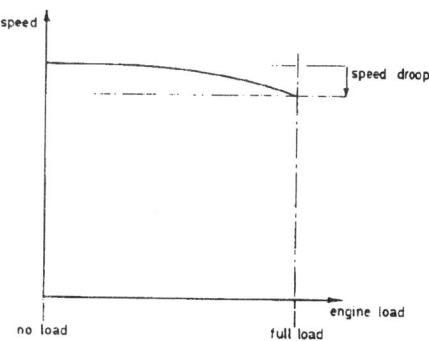

Forces involved in the flyweight governor movement are, inertia, friction and spring. Considerable effort may therefore be required to cause movement, this would necessitate a change in speed without any alteration in governor position. This is bad control, the system is insensitive and various equilibrium speeds are possible. For simple systems these various equilibrium speeds are not an embarrassment, but if we require a better controlled system the two functions that the flyweight governor has to perform would be separated into (1) a speed measuring device and (2) a servo-power amplifier.

Fig. 106. shows the basic arrangement in block form. A load increase would cause a momentary speed droop. The speed measuring device would obtain a measured value signal from the diesel and compare this with a desired value from the speed setting control. The deviation would be converted into an output that would bring into action the servo-power amplifier which would position the fuel rack, increasing the supply of fuel to meet the increase in load.

Since the speed measuring device does not have to position the fuel rack – in fact it could be near zero loaded – it can be very responsive, minimising the time delay between load alteration and fuel alteration in the closed loop. The servo-power amplifier is usually a hydraulic device that simply, quickly and effectively provides the necessary muscle to move the fuel rack.

A proportional action governor is diagrammatically shown in Fig. 107. The centrifugal speed measuring unit is fitted with a conically shaped spring, unlike that shown in Fig. 103., this gives a spring rate which varies as the square of the speed. Fig. 108. This gives linearity to the speed measuring system, *i.e.* the response is directly proportional to the change in speed. If we consider an increase in load on the engine, the pilot valve will move down due to the speed drop. The piston in the servo-amplifier will move up and increase the fuel supply to the engine. The feedback link reduces the force in the speeder spring so that the flyweights can move outwards to a new position, thus raising the pilot valve and closing off the oil supply. If for some reason the oil supply system should fail then the spring loaded piston in the servo-cylinder would be moved down and fuel to the engine cut off. This is called fail safe. Any oil that leaks past the servo-piston will be drained off

CONTROL

FIG 106
BASIC ARRANGEMENT

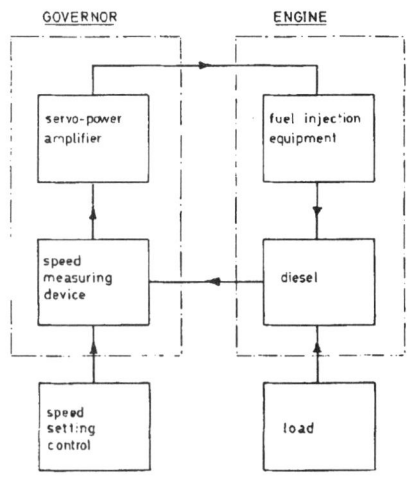

FIG 107
PROPORTIONAL ACTION GOVERNOR

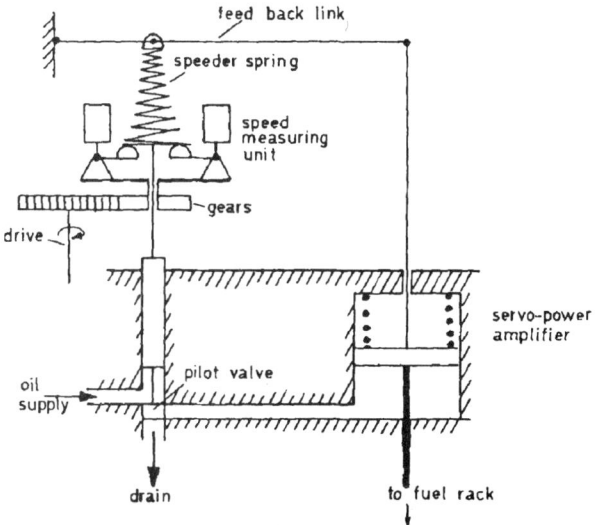

FIG 108
GOVERNOR (SPEED)

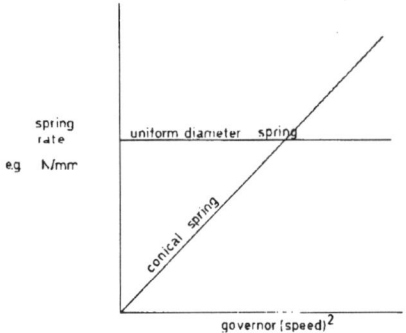

to the oil sump tank. If this were not so the servo-piston would eventually lock in position.

Flywheels and Their Effect

Flywheel dimensions are dictated by allowable speed variation due to non-uniform torque caused by individual cylinders firing. This of course is outside the control of the governor. If the speed has to remain nearly constant during changes of load it may be decided to fit a large flywheel, which increases the moment of inertia of the system and gives an integral effect – this must not be taken to extremes or instability may occur. Flywheels, however, are not cheap and a less expensive solution to the problem may be to fit a better governor.

Integral effect or as it is often called "reset action" reduces offset to zero, *i.e.* during load alteration the speed will go from the desired value but the reset action worked to return the speed to the desired value, so that after the load change the speed is the same as before.

Governor With Proportional and Reset Action

Fig. 109. shows diagrammatically the type of governor that will, after an alteration in engine load, return the speed of the engine back to the value it was operating at before the alteration. If an increase in engine load is considered, the flyweights will move radially inwards and the pilot valve will open to admit oil to the servo-piston. The servo-piston will move up the cylinder

FIG 109
GOVERNOR WITH PROPORTIONAL AND RESET ACTION

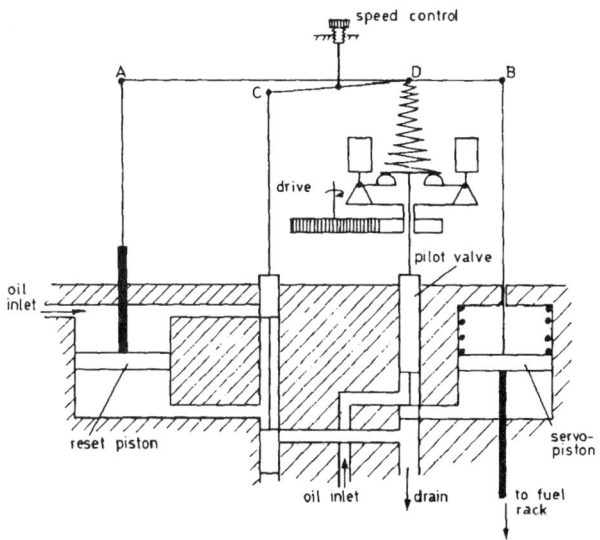

compressing the spring and at the same time it will cause (1) the fuel rack to be repositioned to increase fuel supply to the engine, (2) rotate the feedback link "A-B" anti-clockwise about the pivot point "A" (This point "A" would initially be locked due to equal pressures on either side of the reset piston), (3) rotate link "C-D" will move the reset piston control valve down and some oil will drain from the reset piston cylinder. As the reset piston moves down to a new equilibrium position the feedback link "A-B' will pivot about "B" and the link "C-D" will be rotated clockwise, closing the drain from the reset piston cylinder (and thus locking the reset piston in a new position), returning the point "D" to its original position. This means that the engine is now running at its original speed but with increased fuel supply. Speed droop that took place during the change of the relative positions of the two pistons was transient. This type of governor, that has proportional and reset action, is called in governor parlance an "isochronous governor".

FIG 110
ELECTRIC GOVERNOR

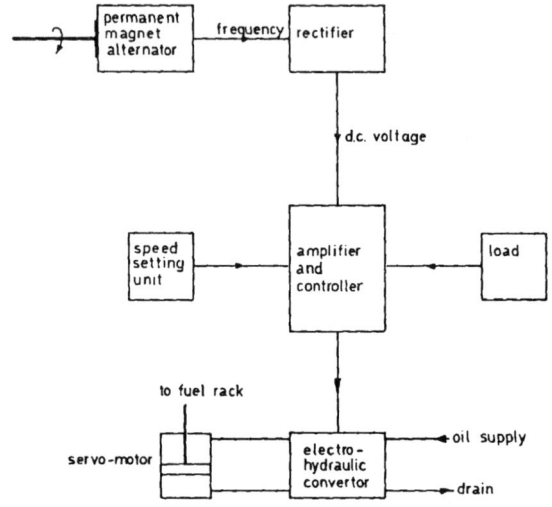

Electric Governor

This governor has proportional and reset action with the addition of load sensing. A small permanent magnet alternator is used to obtain the speed signal, the advantage to be gained is that there will be no slip rings or brushes with their attendant wear. The speed signal obtained from the frequency of the generated a.c. voltage impulses is converted into d.c. voltage which is proportional to the speed. A reference d.c. voltage of opposite polarity, which is representative of the desired operating speed, is fed into the controller from the speed setting unit. These two voltages are connected to the input of an electric amplifier. If the two voltages are equal and opposite, they cancel and there will be no change in amplifier voltage output. If they are different, then the amplifier will send a signal through the controller to the electro-hydraulic converter which will in turn, via the servo-motor, reposition the fuel rack. In order that the system be isochronous the amplifier controller has internal feedback.

Load Sensing

The purpose of including load sensing into the governor is to correct the fuel supply to the prime mover before a speed change occurs. Load sensing governors are therefore anticipatory governors, *i.e.* they anticipate a change in speed and take steps to prevent, as far as possible, its occurrence.

Load sensing could be achieved by mechanical means but it would be a complicated and relatively costly system. For this reason load sensing governors tend to be of the electronic type. The output of, for example, a main generator would be monitored and if a load alteration took place a signal fed to the governor. It must be remembered that the speed of response of the load sensing element must be better than that of the speed sensing element. The speed sensing element would be used to correct small errors of fuel rack position.

In addition to load sensing, electronic governors claim several other advantages:

1. Electronic governors generally have faster response
2. Electronic governors can be mounted in positions remote from the engine thereby eliminating the need for governor drives.
3. Controls and indicators available from electronic governors make automation easier.
4. Control functions, for example, fuel limitation, acceleration and deceleration schedules and shut down functions such as low lubricating oil pressure can be built into the electronic governor.

Geared Diesels

Two diesels geared together must run at the same speed, but if the governors of the two are not set equally then they will not carry equal shares of the load.

In Fig. 111. is shown the governor droop curves for two diesels A and B. Governor A has a higher speed setting than that of B, but since they must both run at a common speed the load carried by A will be greater than that of B. Actual load carried is given by the intersection of the common speed line and the droop curves. By adjusting the speed settings both droop curves could be made to coincide at the intended load, although this would be difficult to achieve in practice.

Shown in Fig. 112. are two sets of droop curves with the same difference in speed settings but with different amounts of speed

FIG 111
LOAD SHARING BETWEEN TWO ENGINES

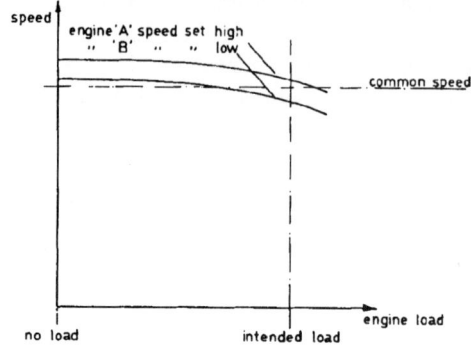

droop. The différence in load sharing at the common speed is less for the larger speed droop curves than for the smaller. Hence speed droop and fine control over the desired level of speed are necessary for effective load sharing.

FIG 112
LOAD DIFFERENCE

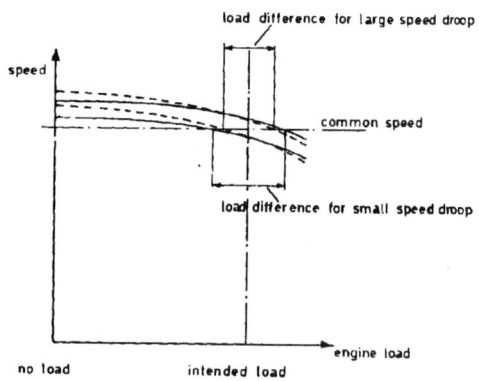

Bridge Control of Direct Drive Diesel Engine

Two consoles would be provided, one on the bridge the other in the engine room. For the bridge console the minimum possible alarms and instruments would be provided commensurate with safety and information requirements, *e.g.* low starting air pressure and temperature, sufficient fuel oil, fuel oil pressure and temperature, etc. The engine room console would give comprehensive coverage and overriding control over that of the bridge.

In Fig. 113. for simplification all normal protective devices are assumed and subsidiary control loops are not considered. The selector would be in the engine room console and the operator can

**FIG 113
ENGINE CONTROL PROGRAMME**

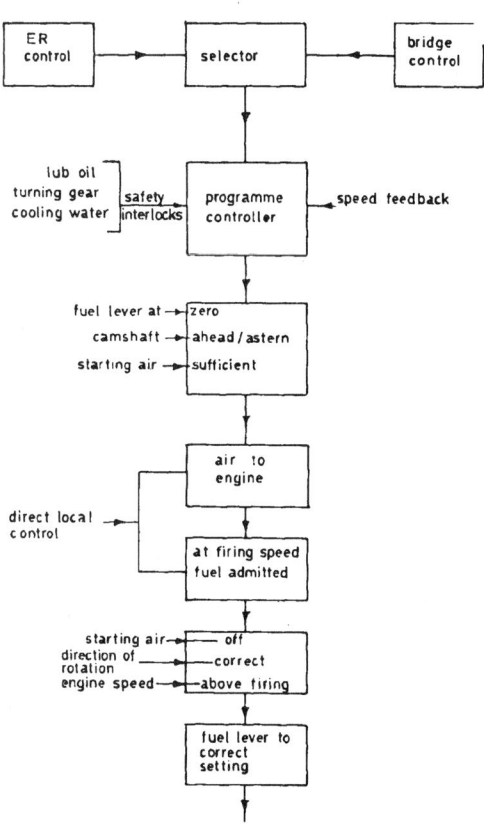

select either engine room or bridge control, with one selected the other is inoperative. Assuming bridge control a programme would be selected, say half ahead, Then providing all safety blockages such as no action with turning gear in, etc. are satisfied, the programme can be initiated and could follow a sequence of checks and operations such as:
 1. Fuel control lever at zero.
 2. Camshaft in ahead position.
 3. Sufficient starting air.
 4. Starting air admitted.
 5. Adjustable time delay permits engine to reach firing speed.
 6. Fuel admitted.
 7. Starting air off, checks on direction of rotation and speed.
 8. Fuel adjusted to set value.

Essential safety locks, such as low lubricating oil pressure or cooling water pressure override the programme and will stop the engine at the same time as they give warning.

Direct local control at the engine itself can be used if required on in the event of an emergency.

Further protective considerations:
 1. Governor, including overspeed trip.
 2. Non operation of air lever during direction alteration.
 3. Failure to fire requires alarm indication and sequence repeat with a maximum of say four consecutive attempts before overall lock.
 4. Movement of control lever for fuel for a speed out of a critical speed range if the bridge speed selection within this range.
 5. Emergency full ahead to full astern timing and setting.

Outline Description

The following is a brief description of one type of electronic-pneumatic bridge control for a given large single screw direct coupled I.C. engine to illustrate the main essentials. The I.C. engine lends itself to remote control more easily than turbine machinery.

Movement of the telegraph lever actuates a variable transformer so giving signals to the engine room electronic controller which transmits, in the correct sequence, a signal series to operate solenoid valves at the engine. One set of solenoid valves controls starting air to the engine while a second set regulates fuel supply, the latter via the manual fuel admission lever, is coupled to a pneumatic cylinder whose speed of travel is governed by an integral hydraulic cylinder in which rate of oil displacement is governed by flow regulators. This cylinder also actuates a variable transformer giving a reset signal when fuel lever position matches telegraph setting.

With the engine on bridge control the engine control box starting air lever is ineffective and the fuel control rack is held clear of the box fuel lever. Engine override of bridge control is provided.

The function of the electronic controller is to give the following sequence for, say, start to half ahead: Ensure fuel at zero, admit starting air in correct direction, check direction, time delay to allow engine to reach firing speed, admit fuel, time delay to cut off air, time delay and check revolutions, adjust revolutions. Similar functions apply for astern or movements from ahead to astern directly. Lever travel time to full can be varied from stop to full between adjustable time limits of 1/2 minute and 6 minutes. Fault and alarm circuits and protection are built into the system.

PISTON COOLING AND LUBRICATING OIL CONTROL

Simple single element control loops can be used for most of the diesel engine auxiliary supply and cooling loops, however during the manoeuvring of diesel engines considerable thermal changes take place with variable time lags which the single element control may not be able to cope with effectively. (N.B. a single element control system is one in which there is only one measuring element feeding information back to the controller.)

For piston cooling and lubricating oil control the use of a cascade control system caters effectively for manoeuvring and steady state conditions. Cascade control means that one controller (the master) is being used to adjust automatically as required the set value of another controller (the slave).

In Fig. 114. the two main variables to consider are sea water inlet temperature and engine thermal load. For simplicity we can consider each variable separately:

1. Assuming the engine thermal load is constant and the sea water temperature varies. The slave controller senses the change in

FIG 114
COOLING AND LUB OIL CONTROL

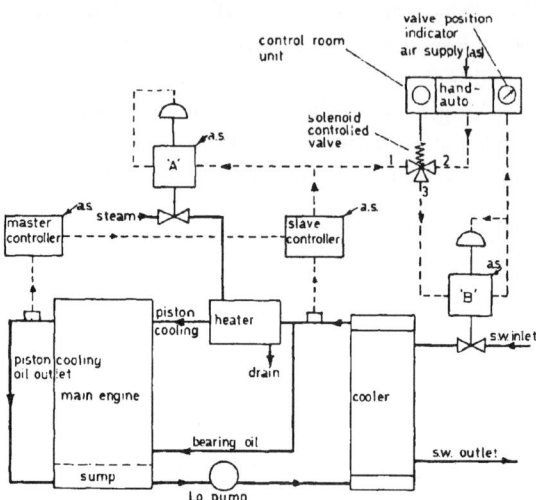

lub. oil outlet temperature from the cooler and compares this with its set value, it then sends a signal to the valve positioner "B" to alter the sea water flow.

2. Assuming the sea water temperature is constant and the engine thermal load falls. The master controller senses a fall in piston cooling oil outlet temperature and compares this with its set value. It then sends a signal to the valve positioner "B" so that the salt water flow will be reduced and the lub.oil temperature at inlet to the piston increased.

If the engine thermal load is low or zero then valve positioner "A" will receive a signal from the slave controller which will cause steam to be supplied to the lub. oil heater. This means that the slave control is split between valve positioners "A" and "B" – this is called "split range control" or "split level control".

Slave controller output range is 1·2 to 2·0 bar.
Valve positioner "A" works on the range 1·2 to 1·4 bar.
Valve positioner "B" works on the range 1·4 to 2·0 bar.
Hence the range is split in the ratio 1·3.

Since the piston cooling oil outlet temperature could be offset from the desired value by upwards of 8°C or more, the master controller must give proportional and reset action. In order to limit the variety of spares that must be carried the slave controller would be identical to the master controller.

It may be necessary to change over from automatic to remote control. This is achieved by position control of the three way solenoid operated valve and regulation of the air supply to the valve positioner "B" at the control room unit. The solenoid operated valve would be positioned to communicate air lines 2 and 3, closing off 1.

Hand regulation of the supply air pressure to valve positioner "B" enables the operator to control the sea water flow to the cooler. Position of the sea water inlet control valve is fed back to control room unit. Lubricating oil temperatures would be indicated on the console in the control room.

An alternative and often preferred arrangement, using a single measuring element, is to have full flow of sea water through the cooler and operate a three way valve (2 inlets, 1 outlet) in the engine fresh water, or cooling circuit that by-passes the cooler.

FIG 115
JACKET (OR PISTON) TEMPERATURE CONTROL

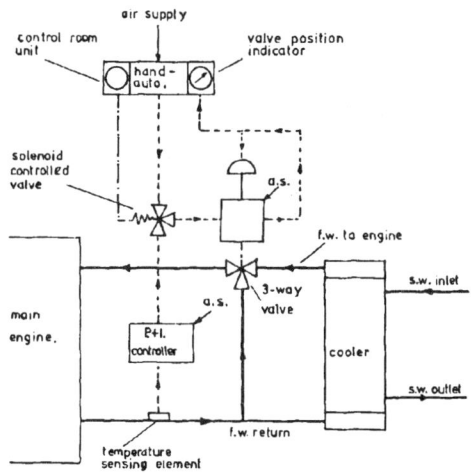

Valve selection for such duties is most important. Maximum pressure and temperature, maximum and minimum flow rate, valve and line pressure drops, etc., must be carefully assessed so that valve selection gives the best results. With correct analysis of the plant parameters and careful valve selection, simple single element control systems can be employed. this would avoid the extra cost of sophisticated control loops and their attendant increased maintenance and fall in reliability.

For mixing and by-pass operations a three-way automatically controlled valve with two inlet and one outlet of the type shown diagrammatically in Fig. 116. could be used.

An increase in controller output pressure p causes the flapper to reduce outflow of air from the nozzle, the pressure on the underside of the diaphragm increases and the valve moves up. As

**FIG 116
3-WAY VALVE AND POSITIONER**

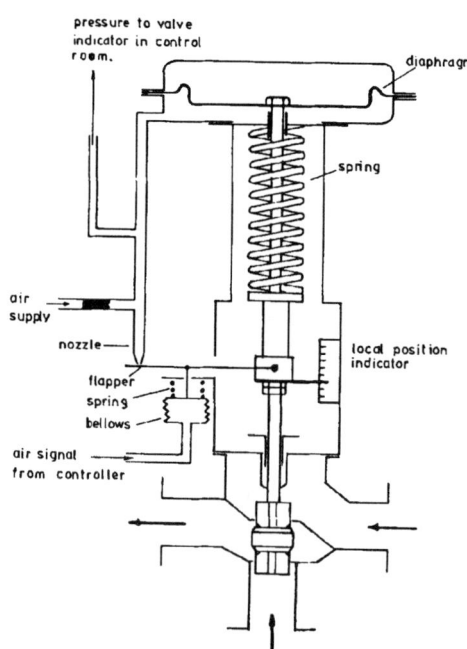

the valve moves, the flapper will be moved to increase outflow of air from the nozzle and eventually the valve will come to rest in a new equilibrium position.

Indication of valve position is given locally and remote, in the latter case by feeding back the diaphragm loading pressure to an indicator possibly situated in the control room. the valve positioner gives accurate positioning of the valve and provides the necessary muscle to operate the valve against the various forces.

Pressure Alarm

The alarm diagrammatically shown in Fig. 117. can be used for either high or low pressure warning. It can also be used for high or low level alarm of fluids in tanks since pressure is a function of head in the tank.

To test the electrical circuitry and freedom of movement of the diaphragm and switches, the hand testing lever can be used. Setting is achieved, for low pressure alarm, by closing the connection valve and opening the drain. When the desired pressure is reached, as indicated on the gauge, the alarm should sound. If high pressure alarm is required the unit can be set by closing the connection valve and coupling a hydraulic pump to the drain connection.

FIG 117
PRESSURE ALARM

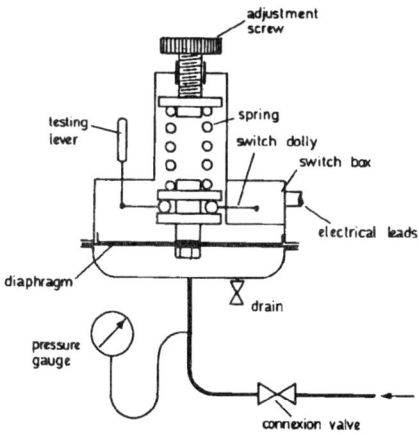

Unattended Machinery Spaces

These are designated u.m.s. in regulations and in the case of the diesel engines they are gradually increasing in number. A controllable pitch propeller driven by geared unidirectional medium or high speed diesels is a relatively uncomplicated system that lends itself to direct control from the bridge. However, irrespective of the type of installation, certain essential requirements for u.m.s. particularly unmanned engine rooms at night, must be fulfilled. They could be summarised as follows:

1. *Bridge control of propulsion machinery*

The bridge watchkeeper must be able to take emergency engine control action. Control and instrumentation must be as simple as possible.

2. *Centralised control and instruments are required in machinery space*

Engineers may be called to the machinery space in emergency and controls must be easily reached and fully comprehensive.

3. *Automatic fire detection system*

Alarm and detection system must operate very rapidly. Numerous well sited and quick response detectors (sensors) must be fitted.

4. *Fire extinguishing system*

In addition to conventional hand extinguishers a control fire station remote from the machinery space is essential. The station must give control of emergency pumps, generators, valves, ventilators, extinguishing media, etc.

5. *Alarm system*

A comprehensive machinery alarm system must be provided for control and accommodation areas.

6. *Automatic bilge high level fluid alarms and pumping units*

Sensing devices in bilges with alarms and hand or automatic pump cut-in devices must be provided.

7. *Automatic start emergency generator*

Such a generator is best connected to separate emergency bus bars. The primary function is to give protection from electrical blackout conditions.

8. *Local hand control of essential machinery.*

9. *Adequate settling tank storage capacity.*

10. *Regular testing and maintenance of instrumentation.*

CHAPTER 7

ANCILLARY SUPPLY SYSTEMS

AIR

Compressed air is used for starting main and auxiliary diesels, operating whistles or typhons, testing pipe lines (*e.g.* CO_2 fire extinguishing system) and for workshop services. The latter could include pneumatic tools and cleaning lances, etc.

Air is composed of mainly 23% Oxygen, 77% Nitrogen by mass and since these are near perfect gases a mixture of them will behave as a near perfect gas, following Boyle's and Charle's laws. When air is compressed its temperature and pressure will increase as its volume is reduced.

Isothermal compression of a gas is compression at constant temperature, this would mean in practice that as the gas is compressed heat would have to be taken from the gas at the same rate as it is being received. This would necessitate a very slow moving piston in a well cooled small bore cylinder.

Adiabatic compression of a gas is compression under constant enthalpy conditions, *i.e.* no heat is given to or taken from the gas through the cylinder walls and all the work done in compressing the gas is stored within it.

In Fig 118. the two compression curves show clearly the extra work done by compressing adiabatically, hence it would be more sensible to compress isothermally. In practice this presents a problem – if the compressor were slow running with a small bore perfectly cooled cylinder and a long stroke piston the air delivery rate would be very low.

Multi-stage compression

If we had an infinite number of stages of compression with coolers in between each stage returning the air to ambient temperature, then we would be able to compress over the desired range under near

FIG 118
A COMPARISON OF COMPRESSION PROCESSES

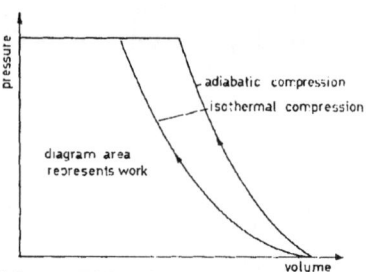

isothermal conditions. This of course is impracticable so two or three stage compression with interstage and cylinder cooling is generally used when relatively high pressures have to be reached.

Fig. 119. shows clearly the work saved by using this method of air compression, but even with efficient cylinder cooling the compression curve is nearer the adiabatic than the isothermal and the faster the delivery rate the more this will be so.

To prevent damage, cylinders have to be water or air cooled and clearance must be provided between piston and cylinder head. This clearance must be as small as practicable.

FIG 119
3 STAGE COMPRESSION

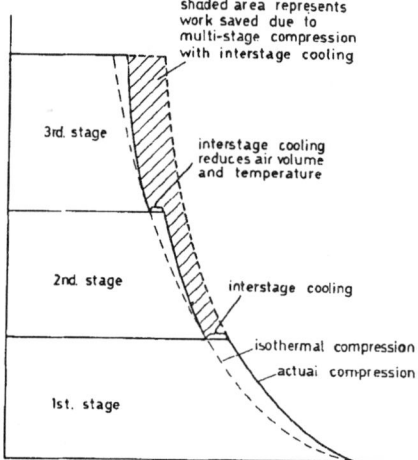

High pressure air remaining in the cylinder after compression and delivery will expand on the return stroke of the piston. This expanding air must fall to a pressure below that in the suction manifold before a fresh air charge can be drawn in. Hence, part of the return or suction stroke of the piston is non-effective. This non-effective part of the suction stroke must be kept as small as possible in order to keep capacity to a maximum.

Volumetric efficiency is a measure of compressor capacity, it is the ratio of the actual volume of air drawn in each suction stroke to the stroke volume. Fig. 120. shows what would happen to the compressor volumetric efficiency – and hence capacity – if the clearance volume were increased.

Clearance volume can be calculated from an indicator card by taking any three points on the compression curve such that their pressures are in geometric progression, *i.e.* $P_1/P_2 = P_2/P_3$ hence $P_2 = \sqrt{P_1 P_3}$. (Fig. 121.). If V_c = clearance volume as a percentage of the readily calculable stroke volume and V_1, V_2, V_3 are also percentages of the stroke volume then:

FIG 120
EFFECTS OF INCREASING CLEARANCE VOLUME

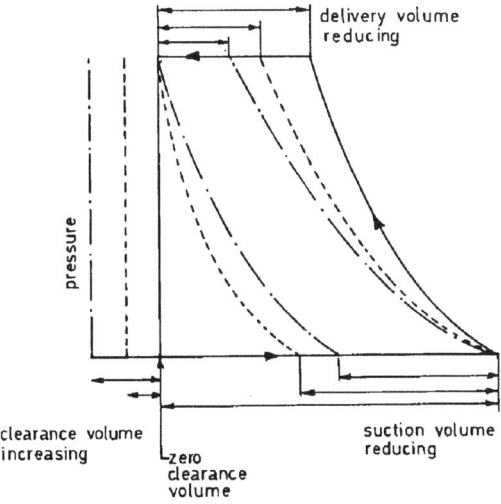

FIG 121
CALCULATING CLEARANCE VOLUME

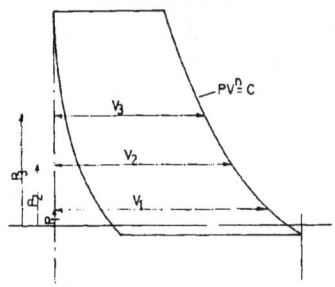

$$P_1(V_1 + V_c)^n = P_2(V_2 + V_c)^n = P_3(V_3 + V_c)^n$$

i.e. $\quad \dfrac{P_1}{P_2} = \left(\dfrac{V_2 + V_c}{V_1 + V_c}\right)^n$ and $\dfrac{P_2}{P_3} = \left(\dfrac{V_3 + V_c}{V_2 + V_c}\right)^n$

now $\quad \dfrac{P_1}{P_2} = \dfrac{P_2}{P_3}$

therefore $\quad \left(\dfrac{V_2 + V_c}{V_1 + V_c}\right)^n = \left(\dfrac{V_3 + V_c}{V_2 + V_c}\right)^n$

hence $\quad \dfrac{V_2 + V_c}{V_1 + V_c} = \dfrac{V_3 + V_c}{V_2 + V_c}$

$$V_c = \dfrac{V_2^2 - V_1 V_3}{V_1 + V_3 - 2V_2}$$

Since V_1, V_2 and V_3 are known V_c can be calculated.

Correct clearance must be maintained and this is usually done by checking the mechanical clearance and adjusting it as required by using inserts under the palm of the connecting rod. Bearing clearances should also be kept at recommended values.

Methods of ascertaining the mechanical clearance in an air compressor:

1. Remove suction or discharge valve assembly from the unit and place a small loose ball of lead wire on the piston edge, then rotate the flywheel by hand to take the piston over top dead centre. Remove and measure the thickness of the lead wire ball.

2. Put crank on top dead centre, slacken or remove bottom half of the bottom end bearing. Rig a clock gauge with one contact touching some underpart of the piston or piston assembly and the other on the crank web. Take a gauge reading. Then by using a suitable lever bump the piston, *i.e.* raise it until it touches the cylinder cover. Take another gauge reading, the difference between the two readings gives the mechanical clearance.

In practice the effective volume drawn in per stroke is further reduced since the pressure in the cylinder on the suction stroke must fall sufficiently below the atmospheric pressure so that the inertia and spring force of the suction valve can be overcome. Fig. 122. shows this effect on the actual indicator card and also the excess pressure above the mean required upon delivery, to overcome delivery valve inertia and spring force.

Air compressors are either reciprocating or rotary types, the former are most commonly used at sea for the production of air for purposes outlined at the beginning of this chapter. The latter types are used to produce large volumes of air at relatively low pressure and are used at sea as integral parts of main engines for scavenging and for boiler forced draught.

Reciprocating air compressors at sea are generally two or three stage types with inter-stage cooling. Fig. 123. shows diagrammatically a tandem type of three stage compressor, the pressures and temperatures at the various points would be roughly as follows:

FIG 122
ACTUAL PRESSURE-VOLUME

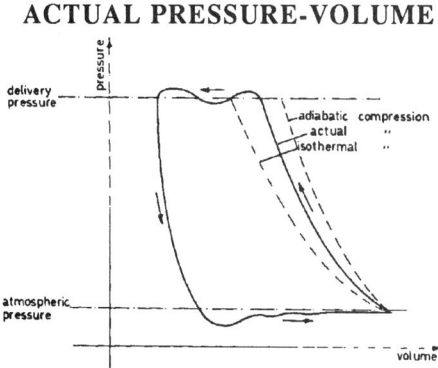

**FIG 123
3 STAGE AIR COMPRESSOR**

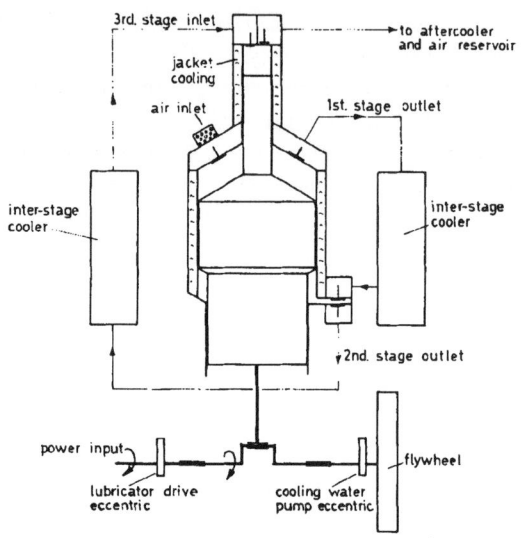

	Delivery pressure	Air temperature	
		Before the coolers	After the coolers
First stage	4 bar	110°C	35°C
Second stage	16 bar	110°C	35°C
Third stage	40 bar	70°C	25°C

The above figures are for a salt water temperature of about 16°C. Final air temperature at exit from the after-cooler is generally at or below atmospheric temperature.

Drains
Fitted after each cooler is a drain valve, these are essential. To emphasise, if we consider 30 m³ of free air relative humidity 75% temperature 20°C being compressed every minute to about 10 bar, about 1/2 litre of water would be obtained each minute.

Drains and valves to air storage unit must be open upon starting up the compressor in order to get rid of accumulated moisture. When the compressor is running drains have to be opened and closed at regular intervals.

ANCILLARY SUPPLY SYSTEMS 201

Filters
Air contains suspended foreign matter, much of which is abrasive. If this is allowed to enter the compressor it will combine with the lubricating oil to form an abrasive-like paste which increases wear on piston rings, liners and valves. It can adhere to the valves and prevent them from closing properly, which in turn can lead to higher discharge temperatures and the formation of what appears to be a carbon deposit on the valves, etc. Strictly, the apparent carbon deposit on valves contains very little carbon from the oil, it is mainly solid matter from the atmosphere.

These carbon like deposits can become extremely hot on valves which are not closing correctly and could act as ignition points for air-oil vapour mixtures, leading to possible fires and explosions in the compressor.

Hence air filters are extremely important, they must be regularly cleaned and where necessary renewed and the compressor must never be run with the air intake filter removed.

Relieving Devices
After each stage of compression a relief valve will normally be fitted. Regulations only require the fitting of a relieving device on the h.p. stage. Bursting discs or some other relieving device are fitted to the water side of coolers so that in the event of a compressed air carrying tube bursting, the sudden rise in pressure of the surrounding water will not fracture the cooler casing. In the event of a failure of a bursting disc a thicker one must not be used as a replacement.

Lubrication
Certain factors govern the choice of lubricant for the cylinders of an air compressor, these are:
  Operating temperature, cylinder pressures and air condition.

Operating Temperature
Affects oil viscosity and deposit formation. If the temperature is high this results in low oil viscosity, very easy oil distribution, low film strength, poor sealing and increased wear. If the temperature is low, oil viscosity would be high, this causes poor distribution, increased fluid friction and power loss.

Cylinder Pressures
If these are high the oil requires to have a high film strength to ensure the maintenance of an adequate oil film between the piston rings and the cylinder walls.

Air Condition
Air contains moisture that can condense out. Straight mineral oils would be washed off surfaces by the moisture and this could lead to excessive wear and possible rusting. To prevent this a compounded oil with a rust inhibitor additive would be used. Compounding agents may be from 5 to 25% of non-mineral oil, which is added to a mineral oil blend. Fatty oils are commonly added to lubricating oil that must lubricate in the presence of water, they form an emulsion which adheres to the surface to be lubricated.

Two Stage Air Compressor
Most modern diesel engines use starting air at a pressure of about 26 bar and to achieve this a two stage type of compressor would be adequate. These compressors are generally of the reciprocating type, with various possible arrangements of the cylinders, or they could be a combination of a rotary first stage followed by a reciprocating high pressure stage. This latter arrangement leads to a compact, high delivery rate compressor.

Fig. 124. shows a typical two-stage reciprocating type of air compressor, the pressures and temperatures at the various points would be approximately as follows:

	Delivery pressure	*Air temperature*	
		Before the coolers	*After the coolers*
First stage	4 bar	130°C	35°C
Second stage	26 bar	130°C	35°C

Compressor Valves
Simple suction and discharge valves are shown in Fig. 125. These would be suitable diagrams for reproduction in an examination. Modern valves are somewhat more streamlined and lighter in order to reduce friction losses and valve inertia. Materials used in the construction are generally:

ANCILLARY SUPPLY SYSTEMS

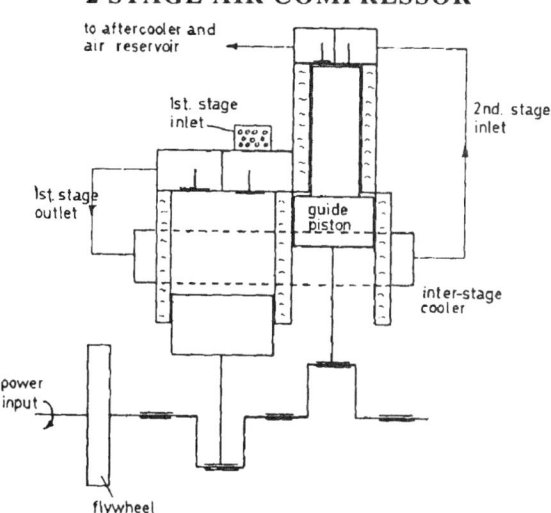

**FIG 124
2 STAGE AIR COMPRESSOR**

Valve Seat
0·4% carbon steel hardened and polished working surfaces.

Valve
Nickel steel, chrome vanadium steel or stainless steel, hardened and ground, then finally polished to a mirror finish.

Spring
Hardened steel. (N.B. all hardened steel would be tempered).

Valve leakages do occur in practice and this leads to loss of efficiency and increase in running time.

Effects of Leaking Valves

1. *First stage Suction*

Reduced air delivery, increased running time and reduced pressure in the suction to the second stage. If the suction valve leaks badly it may completely unload the compressor.

2. *First Stage Delivery*

With high pressure air leaking back into the cylinder less air can be drawn in, this means reduced delivery and increased discharge temperature.

FIG 125
COMPRESSOR VALVES

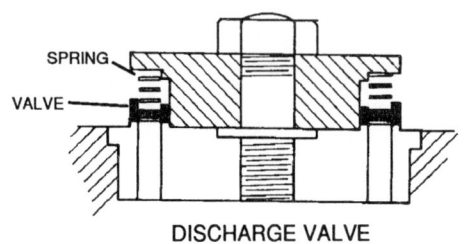

DISCHARGE VALVE

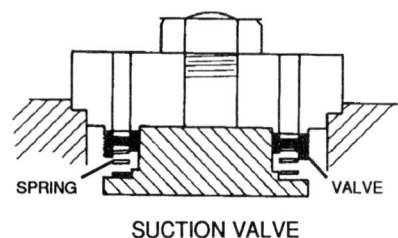

SUCTION VALVE

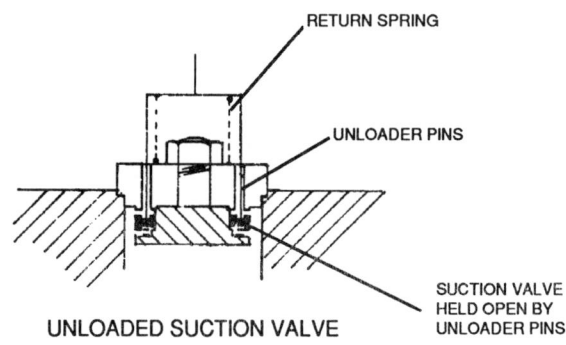

UNLOADED SUCTION VALVE

3. *Second Stage Suction*
High pressure and temperature in the second stage suction line, reduced delivery and increased running time.

4. *Second Stage Delivery*
Increased suction pressure in second stage, reduced air suction and delivery in second stage. Delivery pressure from first stage increased. Fig. 126. shows the effect of a leaking second stage delivery valve on the indicator cards of a compressor.

It must be remembered that it is not usual to find a facility for taking indicator cards from air compressors.

Regulation of Air Compressors
Various methods are available:

Start Stop Control
This is only suitable for small electrically driven types of unit. A pressure transducer attached to the air receiver set for desired max-min pressures would switch the current to the electric motor either on or off. Drainage would have to be automatic and air receiver

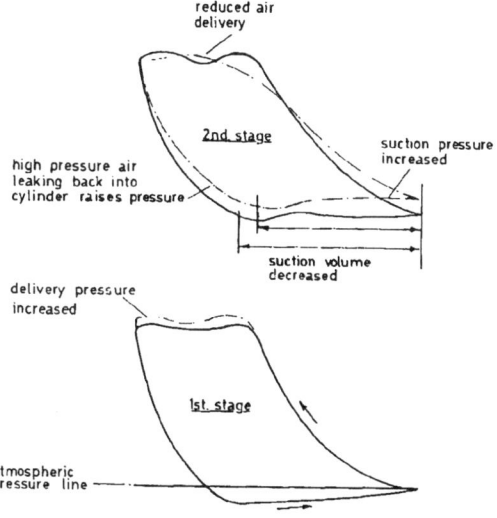

**FIG 126
EFFECT OF LEAKING 2ND STAGE
DELIVERY VALVES**

relatively large compared to the compressor unit requirements so that the number of starts per unit time is not too great. It must be remembered that the starting current for an electric motor is about double the normal running current.

Constant Running Control

This method of control is the one most often used. The compressor runs continuously at a constant speed and when the desired air pressure is reached the air compressor is unloaded in some way so that no air is delivered and practically no work is done in the compressor cylinders.

The methods used for compressor unloading vary, but that most commonly used is on the suction side of the compressor. If the compressor receives no air then it cannot deliver any. Or if the air taken in at the suction is returned to the suction no air will be delivered. In either case virtually no work would be done in the compressor cylinder or cylinders and this would provide an economy compared to discharging high pressure air to the atmosphere through a relief valve.

Fig. 127. shows diagrammatically a compressor unloading valve fitted to the compressor suction. When the discharge air pressure reaches a desired value it will act on the piston causing the spring loaded valve to close shutting off the supply of air to the compressor.

An alternative method of unloading the compressor, while continuing to run it, is to hold the suction valve open. During periods of unload the suction valve plates are held open by pins operated by a relay valve and piston, not unlike that shown in Fig. 127. When the pressure in the air reservoir falls to a preset level, the unloading piston is vented and springs withdraw the pins holding allowing the suction valve to operate normally. Fig. 125.

Automatic Drain

Fig. 128. shows an automatic air drain trap which functions in a near similar way to a steam trap.

With water under pressure at the inlet the disc will lift, allowing the water to flow radially across the disc from A to the outlet B. When the water is discharged and air now flows radially outwards from A across the disc, the air expands increasing in velocity ramming air into C and the space above the disc, causing the disc

FIG 127
COMPRESSOR UNLOADING VALVE

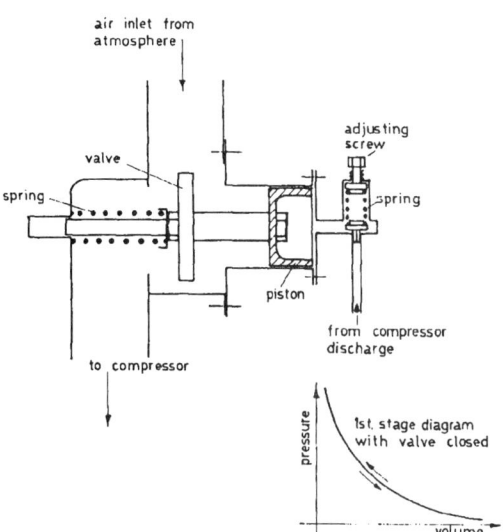

to close on the inlet. Because of the build-up of static pressure in the space above the disc in this way, and the differential area on which the pressures are acting, the disc is held firmly closed. It will remain so unless the pressure in the space above the disc falls.

In order that this pressure can fall, and the trap re-open, a small groove is cut across the face of the disc communicating B and C

FIG 128
AIR DRAIN TRAP

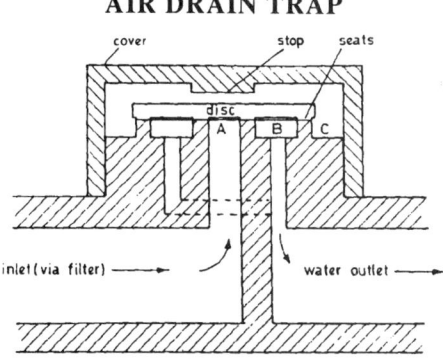

through which the air slowly leaks to outlet.

Obviously this gives an operational frequency to the opening and closing of the disc which is a function of various factors, *e.g.* size of groove, disc thickness, volume of space above the disc, etc. It is therefore essential that the correct trap be fitted to the drainage system to ensure efficient and effective operation.

AIR VESSELS

Material used in the construction must be of good quality low carbon steel similar to that used for boilers, *e.g.* 0·2% Carbon (max.), 0·35% silicon (max.), 0·1% Manganese, 0·05% sulphur (max.), 0·05% Phosphorus (max.), u.t.s. 460 MN/m^2.

Welded construction has superseded the rivetted types and welding must be done to class 1 or class 2 depending upon operating pressure. If above 35 bar approximately then class 1 welding regulations apply.

Some of the main points relating to class 1 welding are that the welding must be radiographed, annealing must be carried out at a temperature of about 600°C and a test piece must be provided for bend, impact and tensile tests together with micrographic and macrographic examination.

Mountings generally provided are shown in Fig. 129. If it is possible for the receiver to be isolated from the safety valve then it must have a fusible plug fitted, melting point approximately 150°C, and if carbon dioxide is used for fire fighting it is recommended that the discharge from the fusible plug be led to the deck. Stop valves on the receiver generally permit slow opening to avoid rapid pressure increases in the piping system, and piping for starting air has to be protected against the possible effects of explosion.

Drains for the removal of accumulated oil and water are fitted to the compressor, filters, separators, receivers and lower parts of pipe-lines.

Before commencing to fill the air vessel after overhaul or examination, ensure:

1. Nothing has been left inside the air vessel, *e.g.* cotton waste that could foul up drains or other outlets.
2. Check pressure gauge against a master gauge.
3. All doors are correctly centred on their joints.

FIG 129
AIR RESERVOIR

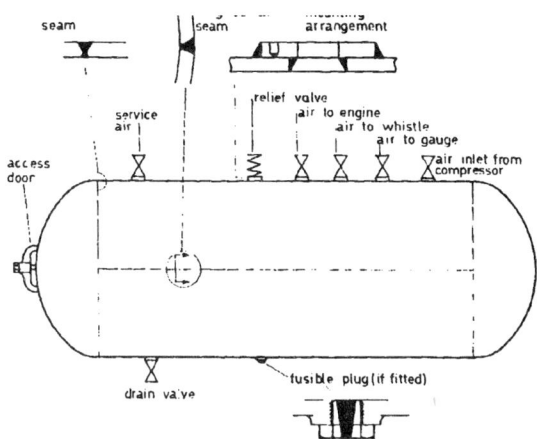

Run the compressor with all drains open to clear the lines of any oil or water, and when filling open drains at regular intervals, observe pressure.

After filling close the air inlet to the bottle, check for leaks and follow up on the door joints.

When emptying the receiver prior to overhaul, etc., ensure that it is isolated from any other interconnected receiver which must, of course, be in a fully charged state.

Cleaning the air receiver internally must be done with caution. any cleaner which gives off toxic, inflammable or noxious fumes should be avoided. A brush down and a coating on the internal surfaces of some protective, harmless to personnel, such as a graphite suspension in water could be used.

COOLING SYSTEMS

These can conveniently be grouped into sections.

1. *Cylinder Cooling*

Or jacket cooling: normally fresh or distilled water. This may incorporate cooling of the turbine or turbines in a turbocharged engine and exhaust valve cooling.

2. *Fuel Valve Cooling*

This would be a separate system using fresh water or a fine mineral oil.

**FIG 130
JACKET COOLING SYSTEM**

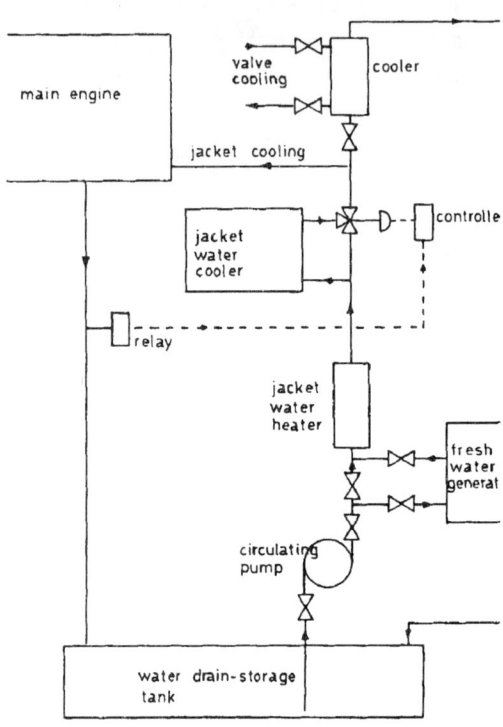

3. *Piston Cooling*

This may be lubricating oil, distilled or fresh water. If it is oil the system is generally common with the lubrication system. If water, a common storage tank with the jacket cooling system would generally be used.

4. *Charge Air Cooling*

This is normally sea water.

Load-Controlled Cylinder Cooling

In an effort to reduce the danger of local liner corrosion over the whole engine load some manufacturers are employing cooling systems that are load dependant. In such a system, shown in Fig. 131., the cooling flow is split into a primary circuit, bypassing the liner, for cylinder head cooling. In the secondary circuit uncooled water from engine outlet is directed to cool the liner. To avoid vapour formation as a result of maintaining higher cooling temperatures the system is pressurised to 4 to 6 bar.

The advantages claimed for such a system include:
1. Possible savings in cylinder lubrication oil feed rate.
2. Omission of cylinder bore insulation.
3. Reduced cylinder liner corrosion.

Comparison Of Coolants
1. Fresh Water

Inexpensive, high specific heat, low viscosity. Contains salts which can deposit, obstruct flow and cause corrosion. Requires treatment. Leakages could contaminate lubricating oil system leading to loss of lubrication, possible overheating of bearings and bearing corrosion. Requires a separate pumping system.

It is important that water should not be changed very often as this can lead to increased deposits. Leakages from the system must be kept to an absolute minimum, so a regular check on the replenishing-expansion tank contents level is necessary.

If the engine has to stand inoperative for a long period and there is a danger of frost, (a) drain the coolant out of the system, (b) heat up the engine room, or (c) circulate system with heating on.

It may become necessary to remove scale from the cooling spaces, the following method could be used. Circulate, with a pump, a dilute hydrochloric acid solution. A hose should be attached to the cooling water outlet pipe to remove gases. Gas emission can be checked by immersing the open end of the hose occasionally into a bucket of water. Keep compartment well ventilated as the gases given off can be dangerous. Acid solution strength in the system can be tested from time to time by putting some on to a piece of lime. When the acid solution still has some strength and no more gas is being given off then the system is scale free. The system should now be drained and flushed out with fresh water, then neutralised with a soda solution and pressure tested to see that the seals do not leak.

FIG 131
LOAD CONTROLLED CYLINDER COOLING

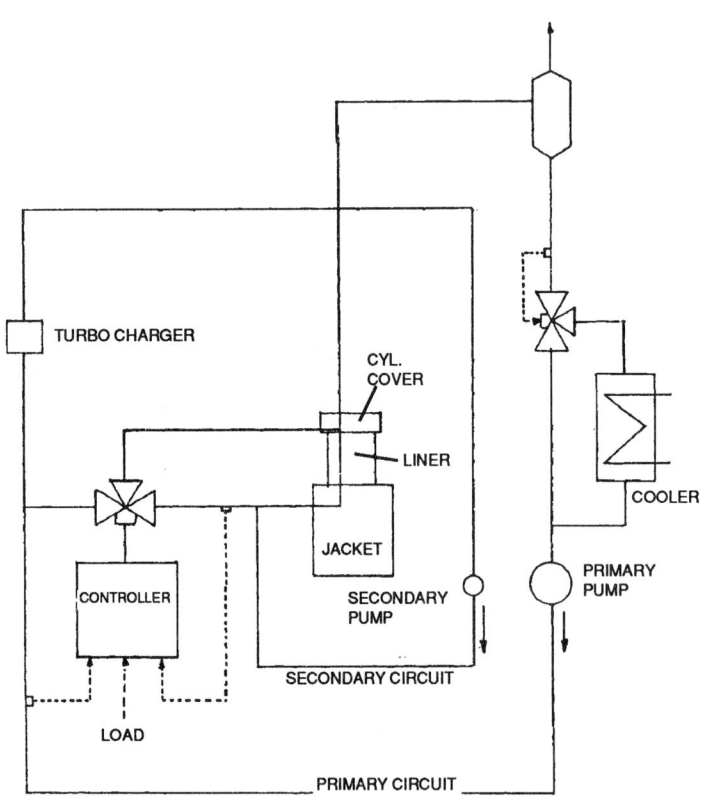

2. Distilled Water

More expensive than fresh water, high specific heat, low viscosity. If produced from evaporated salt water it would be acidic. No scale forming salts. Requires separate pumping system. Leakages could contaminate the lubricating oil system, causing loss of lubrication and possible overheating and failure of bearings, etc.

Additives For Cooling Water

Those generally used are either anti-corrosion oils or inorganic inhibitors.

If pistons are water cooled an anti-corrosion oil is recommended as it lubricates parts which have sliding contact. The oil forms an emulsion and part of the oil builds up a thin unbroken film on metal surfaces, this prevents corrosion but is not thick enough to impair heat transfer.

Inorganic inhibitors form protective layers on metal surfaces guarding them against corrosion.

It is important that the additives used are not harmful if they find their way into drinking water – this is possible if the jacket cooling water is used as a heating medium in a fresh water generator. Emulsion oils and sodium nitrite are both approved additives, but the latter cannot be used if any pipes are galvanised or if any soldered joints exist. Chromates cannot be used if the cooling water is used in a fresh water generator and it is a chemical that must be handled with care.

3. Lubricating Oil

Expensive. Generally no separate pumping system required since the same oil is normally used for lubrication and cooling. Leakages from cooling system to lubrication system are relatively unimportant providing they are not too large, otherwise one piston may be partly deprived of coolant with subsequent overheating

Due to reciprocating action of pistons some relative motion between parts in contact in the coolant supply and return system must occur, oil will lubricate these parts more effectively than water. No chemical treatment required. Lower specific heat than water, hence a greater quantity of oil must be circulated per unit time to give the same cooling effect.

If lubricating oil encounters high temperature it can burn leaving as it does so carbon deposit. This deposit on the underside of a piston crown could lead to impairment of heat transfer, overheating and failure of the metal. Generally the only effective method of dealing with the carbon deposit is to dismantle the piston and physically remove it. Since oil can burn in this way a lower mean outlet and inlet temperature of the oil has to be maintained. In order to achieve this more oil must be circulated per unit time.

Some engines may use completely separate systems for oil cooling of pistons and bearing lubrication, the advantages gained by this method are:

1. Different oils can be used for lubrication and cooling, a very low viscosity mineral oil would be better suited to cooling than lubrication.

2. Additives can be used in the lubricating oil that would be beneficial to lubrication, *e.g.* oiliness agents, e.p. agents and V.I. improvers, etc.

3. Improved control over piston temperatures.

4. If oil loss occurs, then with separate systems the problem of detection is simplified and in the case of total oil loss in either system, the quality to be replaced would not be as great as for a common system.

5. Contamination of the oil in either system may take place. In the event the problem of cleaning or renewal of the oil is not so great.

6. Oxidation of lubricating oil in contact with hot piston surfaces leads to rapid reduction in lubrication properties.

Disadvantages of having two separate systems are: Greater initial cost due to separate storage, additional pipework and pumps. A sealing problem to prevent mixing of the two different oils is created and due to the increased complexity more maintenance would have to be carried out.

FIG 132
LUBRICATING AND COOLING OIL SYSTEM

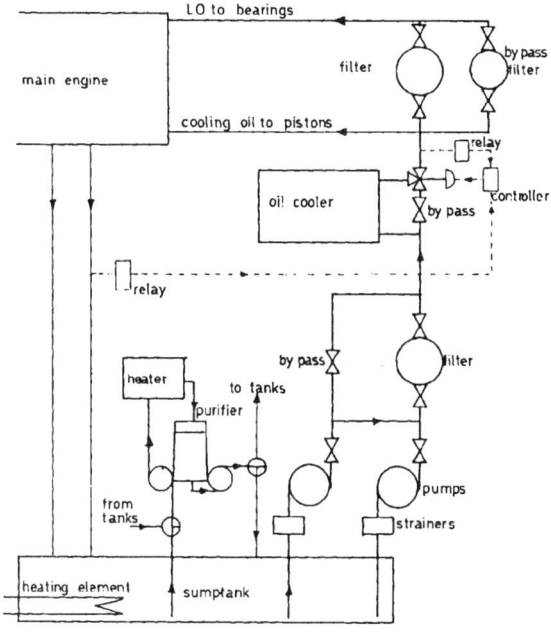

CHAPTER 8

MEDIUM SPEED DIESELS

The term medium speed refers to diesels that operate within the approximate speed range 300 to 800 revolutions per minute. High speed is usually 1000 rev/min and above.

The development in the medium speed engine has been such that it is now a serious competitor for applications which were once only the domain of slow-speed 2-stroke engines. If the advantages and salient features of the medium speed diesel are examined the reader will appreciate why this swing, for certain vessels tonnages, is taking place. They are as follows:

1. Compact and space saving. Vessel can have reduced height and broader beam – useful in some ports where shallow draught is of importance. The considerable reduction in engine height compared to direct drive engines and the reduced weight of components means that lifting tackle, such as the engine room crane, is reduced in size as it will have lighter loads to lift through smaller distances. More cargo space is made available and because of the higher power weight ratio of the engine a greater weight of cargo can be carried.

2. Through using a reduction gear a useful marriage between ideal engine speed and ideal propeller speed can be achieved. For optimum propeller speed hull form and rudder have to be considered, the result is usually a slow turning propeller (for large vessels this can be as low as 50 to 60 rev/min). Gearing enables the Naval Architect to design the best possible propeller for the vessels without having to consider any dictates of the engine.

Engine designers can ignore completely propeller speed and concentrate solely upon producing an engine that will give the best possible power weight ratio.

3. Modern tendency is to utilise uni-directional medium speed geared diesels coupled to either a reverse reduction gear, controllable pitch propeller or electric generator. The first two of these methods are the ones primarily used and the advantages to be

gained are considerable, they are:
  (a) Less starting torque required, clutch disengaged or controllable pitch propeller in neutral.
  (b) Reduced number of engine starts, hence starting air capacity can be greatly reduced and compressor running time minimised. Classification society requirements are six consecutive starts without air replenishment for non-reversible engines and twelve for reversible engines. Cylinder liner wear rate occurs upon starting.
  (c) Engines can be tested at full speed with the vessel alongside a quay without having to take any special precautions.
  (d) With the engine or engines running continuously, power can be taken off via a clutch or clutch/ gear drive for the operation of electric generators or cargo pumps, etc. Hence the engine has become a multi-purpose 'power pack.'
  (e) Improved manoeuvrability, vessel can be brought to rest within a shorter distance by intelligent use of the engines and c.p. propeller.
  (f) Staff load during 'stand-by' periods is reduced, system lends itself ideally to simple bridge control.
4 With two engines coupled via gearing one may be disengaged, whilst the other supplies the motive power, and overhauled. This reduces off hire time and voyage is continued at slightly reduced speed with a fuel saving.
  5. Spare parts are easier to store and manhandle, unit overhaul time will be greatly reduced.

ENGINE COUPLINGS, CLUTCHES AND GEARING

Various arrangements of geared coupled engines are possible, the basic arrangement depends upon the services the engine has to supply, *e.g.* a high electrical load in port may have to be catered for with the alternator being driven at a higher speed than the engine. Hence a step up gear box would be required along with some form of clutch. Large capacity cargo pumps operating at high speed would require a similar arrangement. Fig. 133. shows different types of arrangements with different types of clutches or couplings being used.

FIG 133
ENGINE ARRANGEMENTS

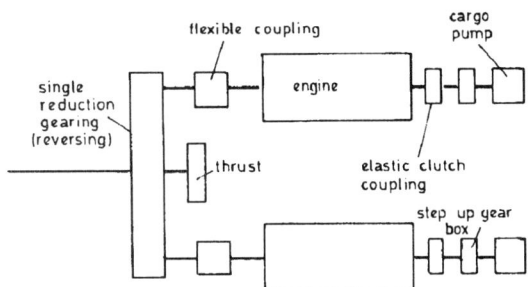

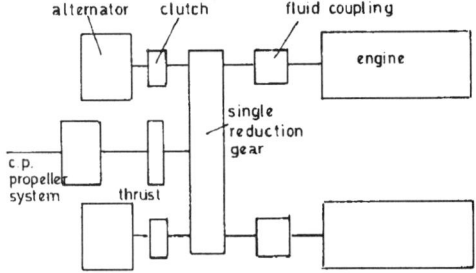

Fluid Couplings

These are completely self-contained, apart from a cooling water supply, they require no external auxiliary pump or oil feed tank. A scoop tube when lowered picks up oil from the rotating casing reservoir and supplies it to the vanes for coupling and power transmission, withdrawal of the scoop tube from the oil stops the flow of oil to the vane which then drains to the reservoir. During power transmission a flow of oil takes place continuously through the cooler and clutch.

Fluid clutches operate smoothly and effectively. They use a fine mineral lubricating oil and have no contact and hence no wear between driving and driven members. Torsional vibrations are dampened out to some extent by the clutch and transmitted speeds can be considerably less than engine speed if required by suitable adjustment of the scoop tube. It is possible to have a dual entry

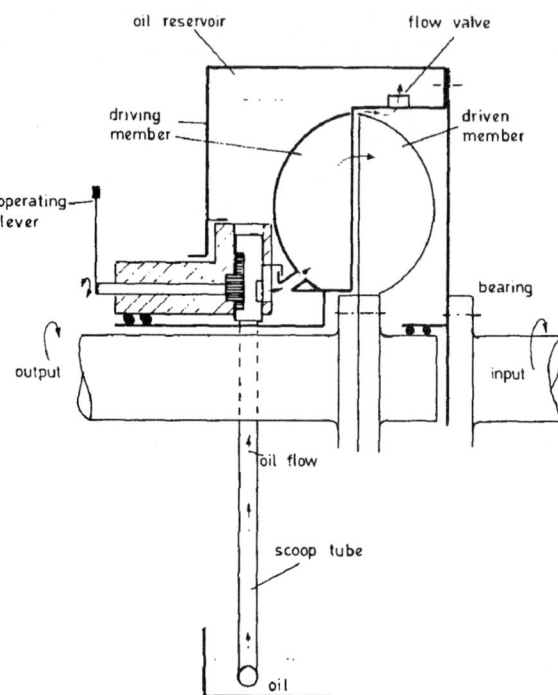

**FIG 134
FLUID COUPLING (VULCAN)**

scoop tube for reversible engines, this obviates the use of c.p. propellers or reversible reduction gears but the control problem is considerably more complex with reversible engines, they have to be stopped and started and if 4-stroke engines are used camshafts have to be moved, etc.

Reverse Reduction Gear

These gear systems are mainly restricted, at present, to powers of up to about 4800kW for twin engined single screw installations. Their obvious advantages are:

    1. Uni-directional engine.
    2. No c.p. propeller required.
    3. Ability to engage or disengage either engine of a twin engine installation from the bridge by a relatively simple remote control.
    4. Improved manoeuvrability, etc.

When dealing with large powers the friction clutches used in the system can become excessively large, great heat generation during engagement may require a cooling system, the whole becomes more expensive and it may be cheaper to use direct reversing engines – however it would for reasons previously outlined be prudent to use a c.p. propeller.

Two systems of reverse reduction gear are shown in Figs. 135. and 136. In Fig. 135. the engine drives a steel drum which has two inflatable synthetic rubber tubes bonded to its inner surface. These tubes have friction material, like brake lining, on their inner surface. Air is supplied through the centrally arranged tube, or the annulus formed by the tube and shaft hole to one or the other of the inflatable tubes. Two flanged wheels are connected via hollow shafts and gears to the main gear wheel and shaft.

For operation ahead, air would be supplied to inflatable tube A. which would then by friction on flanged wheel B. bring gears 1 and 2 up to speed, gears 3, 4 and 5 together with flanged wheel D would be idling.

For astern operation, air would be supplied to inflatable tube C. (A. evacuated) and by friction on flanged wheel D. gears 3, 4, 5

**FIG 135
FRICTION CLUTCH**

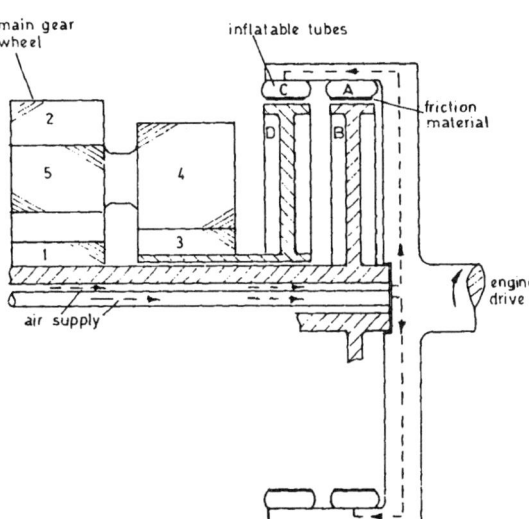

FIG 136
REVERSIBLE REDUCTION GEAR

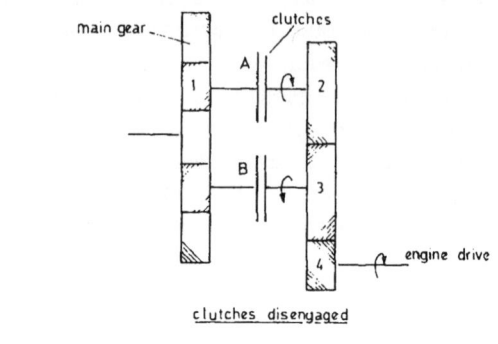

clutches disengaged

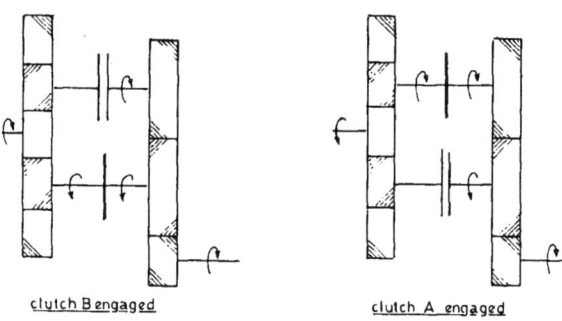

clutch B engaged   clutch A engaged

and 2 would be brought up to speed, gear 1 and drum B. would be idling. For single reduction, gears 3 and 4 would be the same size and so would be gears 1 and 5.

An alternative system, either single or double reduction but probably the latter, is shown in Fig. 136. Friction clutches A. and B. are pneumatically controlled from some remote position. Gears 1, 2, 3 and 4 would have to be the same size if the gear were to be single reduction – but this is most unlikely.

Flexible Couplings

These are used between engine and gearbox to dampen down torque fluctuations, reduce the effects of shock loading on the gears and engine, cater for slight misalignments. They are also used in conjunction with clutches for power take-off when required. In construction they may be similar to the well known multi-tooth type to be found in turbine installations or employ diaphragms or rubber blocks. Those types that use rubber or synthetic rubber, such as Nitrile, give electrical insulation between driving and driven members, but all types will minimise vibration and reduce noise level.

Fig. 137. shows a combination of flexible couplings and pneumatically operated friction clutch, the arrangement gives a smooth transition of speed and torque during engagement, it could be typical of an arrangement for the take off for electrical power or cargo pumps, etc. The rubber blocks would be synthetic if oil is likely to be present as natural rubber is attacked by oil.

The Geislinger Coupling

The main function of a Geislinger coupling is to assist in the damping out of torsional vibrations. This is accomplished by connecting the engine crankshaft to the load via flexible steel leaf springs arranged radially in the coupling, which is also filled with oil. As torsional fluctuations occur they are absorbed by the leaf springs which deflect and displace oil to adjacent chambers, slowing down the relative movement between the inner and outer

**FIG 137
FLEXIBLE CLUTCH COUPLING**

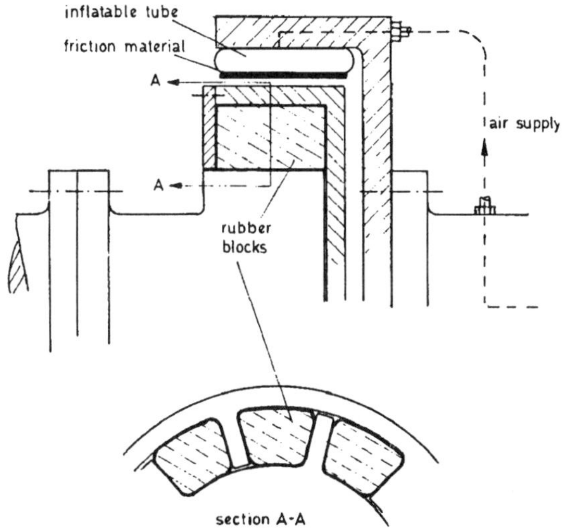

components of the coupling. The makers claim that this effective damping is achieved without problems of wear because of the absence of friction. Fig. 138.

Damping oil is supplied from the engine oil system through the centre of the coupling. It is returned to the engine through hollow coupling bolts. Maintenance is limited to cleaning, inspection and the replacement of "O" rings.

EXHAUST VALVES

Most 4-stroke medium speed diesels incorporate two exhaust valves per cylinder and if we consider a moderate sized installation consisting of two 12-cylinder V engines this gives a total of 48 exhaust valves. A not inconsiderable quantity, and if the plant is to burn fuel of high viscosity, the maintenance problem for these valves could be considerable.

In order to minimise maintenance and to prolong valve life, bearing in mind that burning of high viscosity oil is essential due to the higher cost of light diesel oil, certain design parameters and operating procedures must be followed. These are:

1. Separately caged exhaust valves are preferred even though they increase first cost. If they are made integral with the cylinder head and a fuel of poorer quality than normal has to be burnt, increased frequency of replacement and overhaul of the valves necessitating cylinder head removal each time becomes a tedious time consuming operation.

2. All connections to the valves, cooling, exhaust, etc. should be capable of easy disconnection and re-assembly.

3. Materials that have to operate at elevated temperatures must be capable of withstanding the erosive and corrosive effects of the exhaust gas. When burning oils of high viscosity which contain Sodium and Vanadium deposits can form on the valve seats which, at high temperatures, (in excess of 530°C at the valve seat) become strongly corrosive sticky compounds which lead to burnt valves. Hence the need for materials that can withstand the corrosion and for intense cooling arrangements for valve seats.

Stellited valve seats are not uncommon, Stellite is a mixture of Cobalt, Chromium and Tungsten extremely hard and corrosion resistant that is fused on to the operating surfaces.

FIG 138
GEISLINGER TORSIONAL VIBRATION DAMPING COUPLING

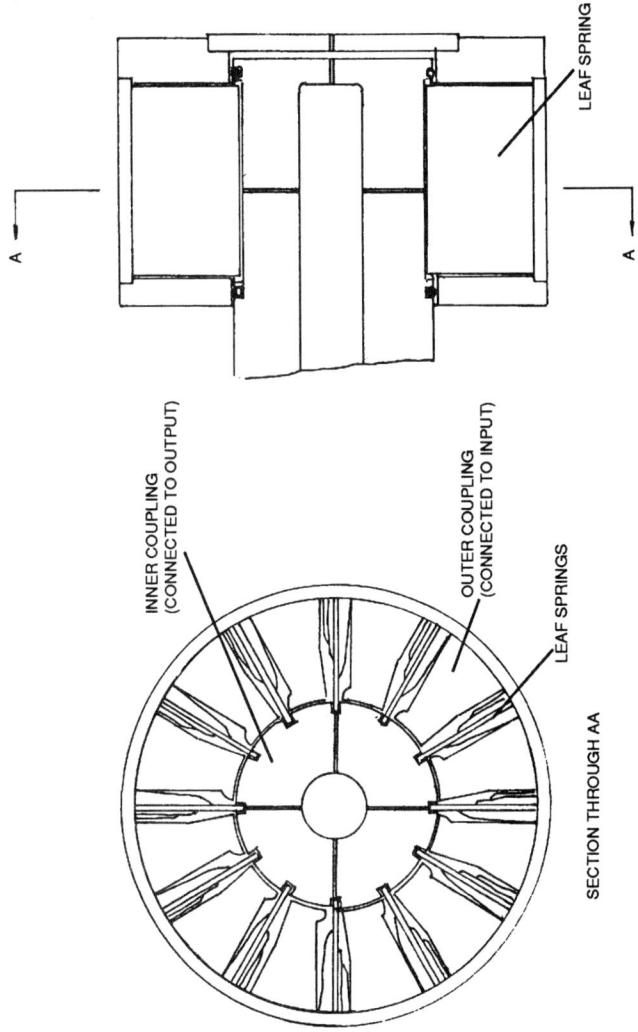

Low temperature corrosion due to sulphur compounds can occur during prolonged periods of running under low load conditions. The valve spindle and guide, which would be at a relatively low temperature, are the principal places of attack due to the effective cooling in this region. Ideally, valve cooling should be a function of engine load with the valve being maintained at a uniform temperature at all times, this could prove complicated and expensive to arrange for.

4. Effective lubrication of the valve spindle is necessary to avoid risk of seizure and possible mechanical damage due to a valve 'hanging up.' In order to minimise lubricating oil usage the lubrication system for the valves would be similar to that used for cylinder lubrication and since the amount of oil used would therefore be in small quantities any contamination of the oil by combustion products and water, etc. would be minimal, this would also increase the life of crankcase lubricating oil.

Rotocap
This simple device when fitted to exhaust valves causes rotation of the valve spindle during valve opening, wear of the valve seat is reduced, seat deposits are loosened, valve operation life is extended. Fig. 139. shows the Rotocap which operates as follows: An increase in spring force on the valve as it opens flattens the belleville washer so that it no longer bears on the bearing housing (B.) at A., this removes the frictional holding force between B. and C., the spring cover. Further increase in spring force causes the balls to move down the ramps in the retainer imparting as they move a torque which rotates the valve spindle. As the valve closes, and load from the belleville washer is removed from the balls and they return to the position shown in section D-D.

Fig. 140. shows an exhaust valve with welded stellited seat around which cooling water flows keeping the metal temperature at full load conditions well below 500°C, minimising the risk of attack by sodium-vanadium compounds. The valve is housed in a "cage" which can be easily removed for maintenance without disturbing the cylinder cover.

It has been stated in chapter 2 that modern medium speed 4-stroke engines usually have 4 valve cylinder heads to maximise the cross-sectional area of the ports and thus improve gas flow through the engine. The gas flow of a typical 4 valve cylinder head is shown in Fig. 141.

FIG 139
ROTOCAP

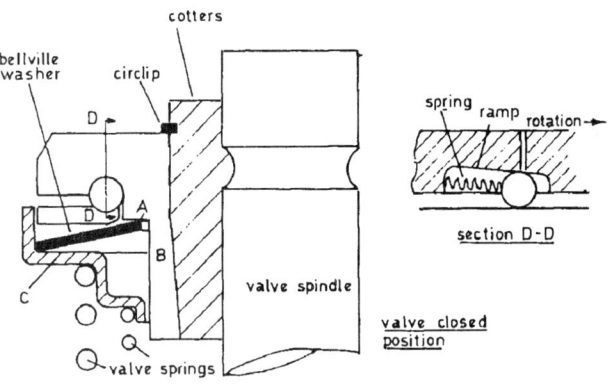

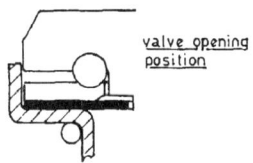

ENGINE DESIGN

The principal design parameters for a medium speed diesel engine are:
1. High power/weight ratio.
2. Simple, strong, compact and space saving.
3. High reliability.
4. Able to burn a wide range of fuels.
5. Easy to maintain, the fact that components are smaller and lighter than those for slow speed diesels makes for easier handling, but accessibility and simple to understand arrangements are inherent features of good design.
6. Easily capable of adaption to unmanned operation.
7. Low fuel and lubricating oil consumption.
8. High thermal efficiency.
9. Low cost and simple to install.

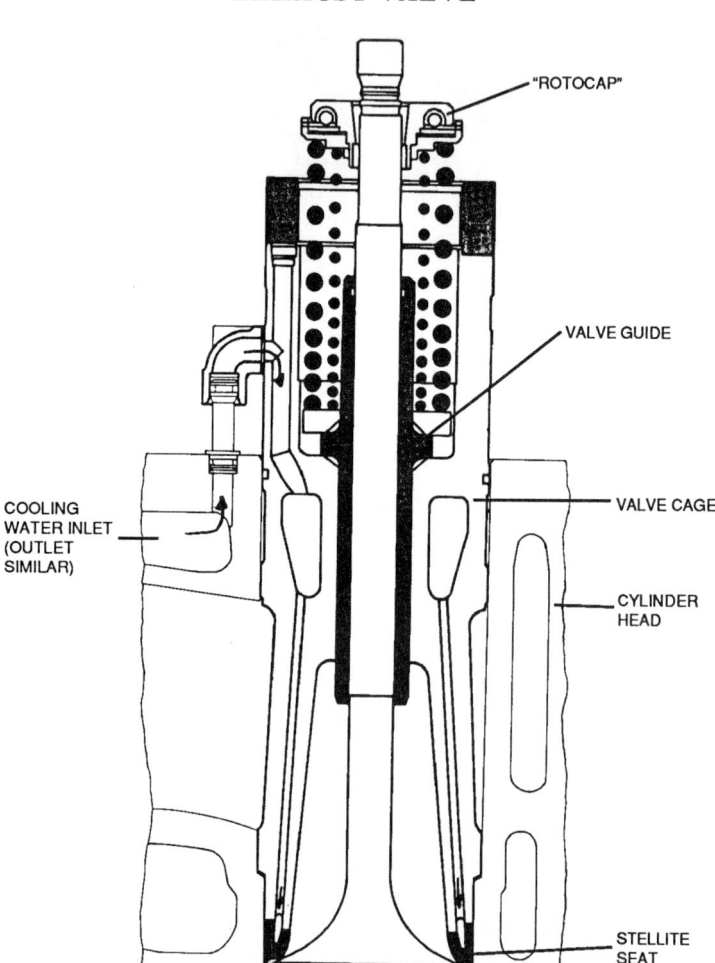

FIG 140
EXHAUST VALVE

Types of Engine

Either 2- or 4-stroke cycle single acting turbocharged with 'in line' or 'V' cylinder configuration. The main choice is, certainly at present, for the 4-stroke engine and there are various reasons for this.

FIG 141
GAS FLOW OF TYPICAL 4 VALVE CYLINDER HEAD

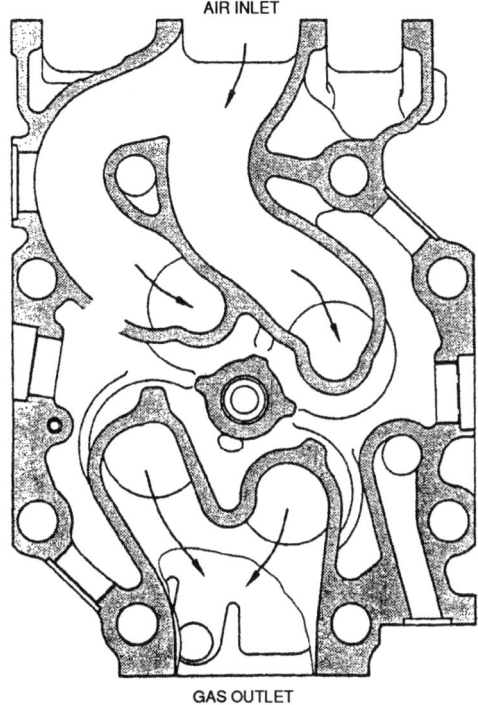

1. They are capable of operating satisfactorily on the same heavy oils as slow speed 2-stroke engines.
2. Effective scavenging is relatively easy to achieve in slow speed 2-stroke engines but it becomes more difficult with an increase in mean piston speed. Modern medium speed engines are generally, but not exclusively, of the 4-stroke configuration. With large inlet and exhaust valve overlap effective scavenging can be accomplished. Scavenging is further improved by utilising high turbocharger pressure ratios. Pressure ratios of 3·5 to 4·0 are not now uncommon with 4·5 to 5·0 being developed. Good scavenging and high turbocharger pressure ratio results in engines producing high turbocharger pressure ratio results in engines producing high BMEP figures.

3. Higher mean piston speed. Mean piston speed is simply twice the stroke times the revs/second. For medium speed diesels it would be approximately 9 to 10 m/s and for slow running diesels 7 to 9 m/s would be about average.

In order that greater power can be developed in the cylinder the working fluid must be passed through faster, hence the higher the mean piston speed for a given unit the greater the power. However, practical limitations govern the piston speed. The relation between cylinder cross sectional area and areas of exhaust and air inlet, method of turbocharging and inertia forces are the main limitations.

To reduce inertia forces designers have in the past utilised aluminium alloy for piston skirts and in some cases entire pistons. However, as the output of medium speed engines have increased the limitations of aluminium become apparent. Designers of high output engines now specify cast or forged steel for piston crowns and nodular cast iron for piston skirts. The greater mass of this type of piston means that the higher inertia forces result and cognisance of this must be made when designing the connecting rod and bottom end arrangements. Inertia forces must be taken into account for bearing loads – important in trunk piston engines (*i.e.* the majority of medium and high speed diesels) where the guide surface is the cylinder liner, the smaller the side thrust the less the friction and wear.

4. Engine can operate effectively with the turbocharger out of commission, this would present a considerable problem with some 2-stroke engines of the medium speed type.

5. Turbocharger size and power can be reduced.

6. It is also claimed that the fuel consumption would be reduced.

Typical 'V' Type Engine

The following is a brief description of a medium speed diesel engine currently in use:
- Cylinder bore 400 mm.
- Stroke 560 mm.
- BMEP 23 bar
- Maximum cylinder pressure 160 bar
- 4-stroke turbocharged with up to 18 cylinders developing approximately 700 kW [MCR] per cylinder at approximately 600 rev/min.

Overall dimensions of a 18 cylinder 'V' type
- Length 10·25 m.
- Height 5·0 m.
- Width 4·0 m.
- Dry weight 145 Tonnes.
- Specific fuel consumption 175 g/kW hr

Bedplate and cylinder blocks are of heavy section cast iron, this gives a strong compact arrangement with good properties for damping out vibrations.

The crankshaft, of an "underslung" design, is a solid forging. The connecting rod is also forged but is of the "marine-type" bottom end and is two pieces.

Pistons are of a composite design with forged steel crown and a cast iron skirt. Piston crown is bore cooled.

Liners are of good quality grey cast iron alloy and are bore cooled in the vicinity of the combustion space.

Future Development
The trend in the field of the medium speed engine is towards higher power outputs per cylinder, with high reliability, when operating on cheaper high viscosity fuels. Much development work is being carried out by manufacturers to improve the combustion process. This work focuses on the timing and duration of fuel injection to achieve reliable combustion and manufacturers are now testing engines operating with firing pressures in excess of 210 bar.
- Cylinder bore 580 mm.
- Stroke 600 mm
- Speed 450 rev/min
- Power per cylinder 1250 kW.

Typical Lubrication and Piston Cooling System
A pump, which could be main engine driven, supplies oil to a main feeder pipe wherein oil pressure is maintained at approximately 6 bar. Individual pipes supply oil to the main bearings from the feeder, the oil then passes through the drilled crankshaft to the crank pin bearing then flows up the drilled connecting rod to lubricate the small end bush. It then flows around the cooling tubes cast in the piston crown then back down the connecting rod to the

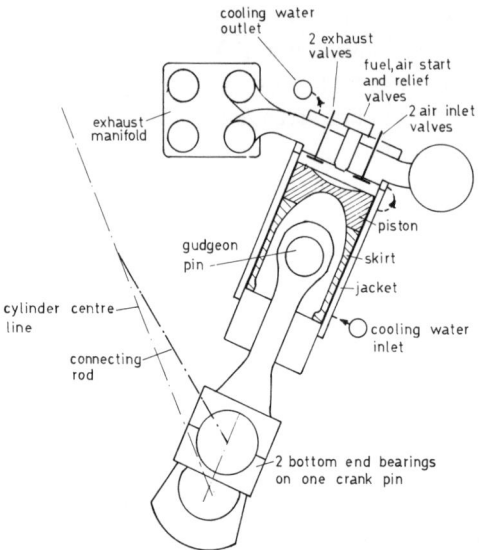

**FIG 142
V-TYPE ENGINE**

engine sump. Oil would also be taken from the main feeder to lubricate camshaft gear drive, camshaft bearings, pump bearings, etc.

Fig. 143. shows in simplified form a typical cooling system for alloy pistons, cast in the piston is a cooling coil and a cast iron ring carrier (marked 1 in the diagram). (2) are two chromium plated compression rings, (3) two copper plated compression rings, (4) two spring backed downward scraping, scraper rings of low inertia type. They are spring backed to give effective outward radial pressure since the gas pressure behind the ring would be very small. The oil flow direction tube is expanded at each end into the gudgeon pin and it is so passaged to direct oil flow and return to their respective places without mixing.

Due to complex vibration problems that can arise in medium speed engines of the 'V' type it would appear important to have a very strong and compact arrangement of bedplate etc. Excessive vibration of the structure can lead to increased cylinder liner wear and considerable amounts of lubricating oil being consumed.

FIG 143
PISTON COOLING

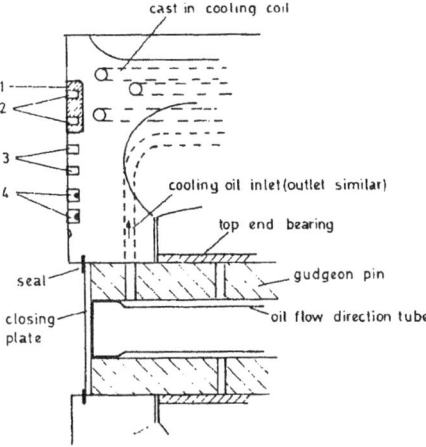

Alkaline lubricating oil of the type used in these engines is expensive and because the engines are mainly trunk type consumption rates can be high. Positioning, and type, of oil scraper ring is important. With some engines they have been moved from a position below the gudgeon pin to above since considerable end leakage sometimes occurred from the gudgeon bearing. The rings should scrape downwards and there may be two scraper rings fitted each with two downward scraping edges, spring backed and of low inertia.

FIG 144
VARIATIONS OF CONNECTING ROD DESIGN

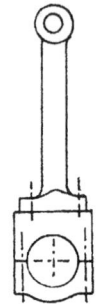 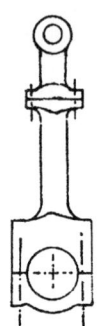

"MARINE TYPE" CONNECTING ROD

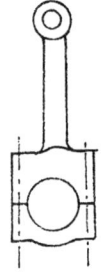 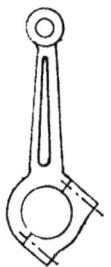 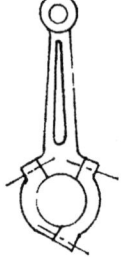

CONNECTING RODS MAY BE ROUND OR H SECTION

CHAPTER 9

WASTE HEAT PLANT

General Details

Reference should be made to chapter 1 for general comments relating to heat balance. Fig. 3 details an approximate heat balance for an IC engine showing significant losses to the exhaust and cooling. Every attempt must be made to utilise energy in waste heat and recovery from both exhaust and coolant is established practice. Sufficient energy potential can be available in exhaust gas at full engine power to generate sufficient steam, in a waste heat boiler, to supply total electrical load and heating services for the ship. The amount of heat actually recovered from the exhaust gases depends upon various factors such as steam pressure, temperature, evaporation rate required, mass flow of gas, condition of heating surfaces, etc. Waste heat boilers can recover up to about 60% of the loss to atmosphere in exhaust gases. Heat recovery from jacket cooling water systems at a temperature of 70-80°C is generally restricted to supplying heat to the fresh water generator.

Combustion Equipment

Obviously most boilers and heaters have arrangements for burning oil fuel during low engine power conditions. It is therefore appropriate to repeat some very general remarks on combustion with details of typical equipment in use.

Good combustion is essential for the efficient running of the boiler as it gives the best possible heat release and the minimum amount of deposits upon the heating surfaces. To ascertain if the combustion is good we measure the % CO_2 content (and in some installations the % O_2 content) and observe the appearance of the gases.

If the % CO_2 content is high (or the % O_2 content low) and the gases are in a non smokey condition then the combustion of the fuel is correct. With a high % CO_2 content the % excess air required for combustion will be low and this results in improved boiler efficiency since less heat is taken from the burning fuel by the small amount of excess air. If the excess air supply is increased then the % CO_2 content of the gases will fall.

Condition of burners, oil condition pressure and temperature, condition of air registers, air supply pressure and temperature are all factors which can influence combustion.

Burners
If these are dirty or the sprayer plates damaged then effective atomisation will not be achieved. Resulting in poor combustion.

Oil
If the oil is dirty it can foul up the burners. (Filters are provided in the oil supply lines to remove most of the dirt particles but filters can get damaged. Ideally the mesh in the last filter should be smaller than the holes in the burner sprayer plate.)

Water in the oil can affect combustion, it could lead to the burners being extinguished and a dangerous situation arising. It could also produce panting which can result in structural defects.

If the oil temperature is too low the oil does not readily atomise since its viscosity will be high, this could cause flame impingement, overheating, tube and refractory failure. If the oil temperature is too high the burner tip becomes too hot and excessive carbon deposits can then be formed on the tip causing spray defects, these could again lead to flame impingement on adjacent refractory and damage could also occur to the air swirlers. Oil pressure is also important since it affects atomisation and lengths of spray jets.

Air Register
Good mixing of the fuel particles with the air is essential, hence the condition of the air registers and their swirling devices are important, if they are damaged mechanically or by corrosion then the air flow will be affected.

Air
The combustion air supply is governed by the combustion controller fuel/air ratio setting. If this is set too low then insufficient air will be supplied resulting in incomplete combustion and the generation of black smoke. If the fuel/air ratio is set too

high then too much air will be supplied for combustion resulting in a greater percentage of free oxygen in the uptakes than is desirable, causing the boiler efficiency to fall.

It is generally considered that the appearance of the boiler uptake gases will give an accurate indication of the effectiveness of combustion. While this is undoubtedly true it should be noted that clear uptake gases can be achieved while supplying excess air, resulting in a reduction in boiler efficiency. To achieve maximum boiler efficiency the fuel/ air ratio setting should be reduced until the setting for optimum combustion, commensurate with clear uptake gases, is reached.

PACKAGE BOILERS

Although such boilers are not necessarily involved with waste heat systems it is considered appropriate to include them at this stage. These boilers are often fitted on motorships for auxiliary use and the principles and practice are a good lead into general boiler practice. Two types of design involving modern principles will now be considered.

Sunrod Vertical Boiler

The design sketched in Fig. 145. is the Sunrod Marine Boiler. This boiler utilises a water-cooled furnace incorporating membrane walled construction. The membrane water wall is backed by low temperature insulation. Fig. 146a. The water wall tubes are joined at the lower end to a circular header and at their upper ends to the steam chamber. Good circulation is assured by the arrangement of a number of downcomers as shown in the diagram. The steam chamber has a number of smoke tubes each fitted with a "Sunrod Element". The purpose of the Sunrod element is to increase the heating surface area of the boiler. This is accomplished by welding pins onto the element as shown in Fig. 146b. In some Sunrod designs the firetube is also water-cooled. This design is manufactured in sizes ranging from 700 kg/hr to 35000 kg/hr with pressures up to 18 bar. The boiler is usually fitted with automatic start up/shut down and combustion control.

Due to the absence of furnace refractory lining this type of boiler is extremely robust and easy to operate. Cleaning the boiler

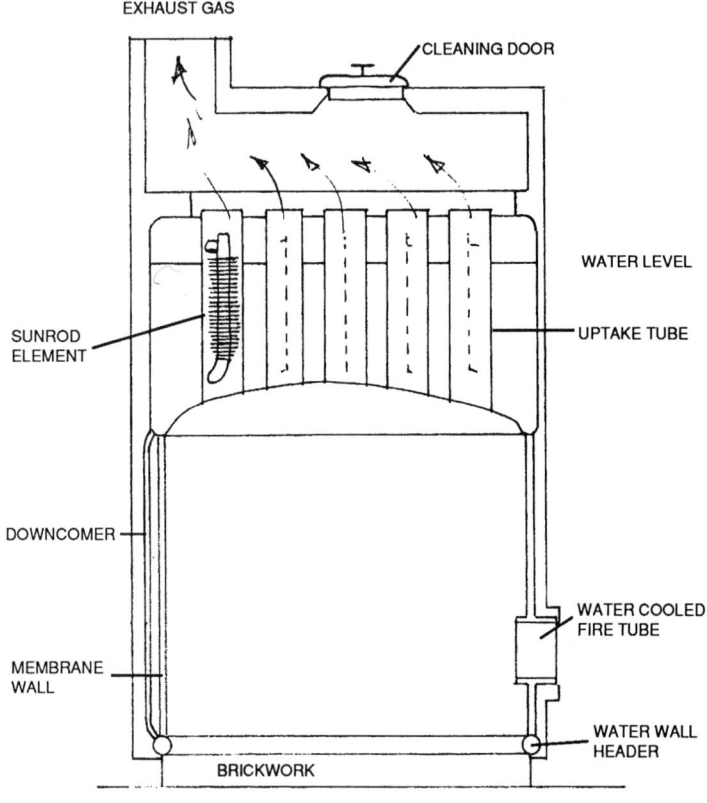

**FIG 145
SUNROD MARINE BOILER**

FIG 146
SUNROD BOILER DETAIL

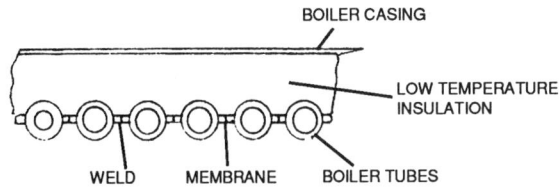

a) MEMBRANE WALL

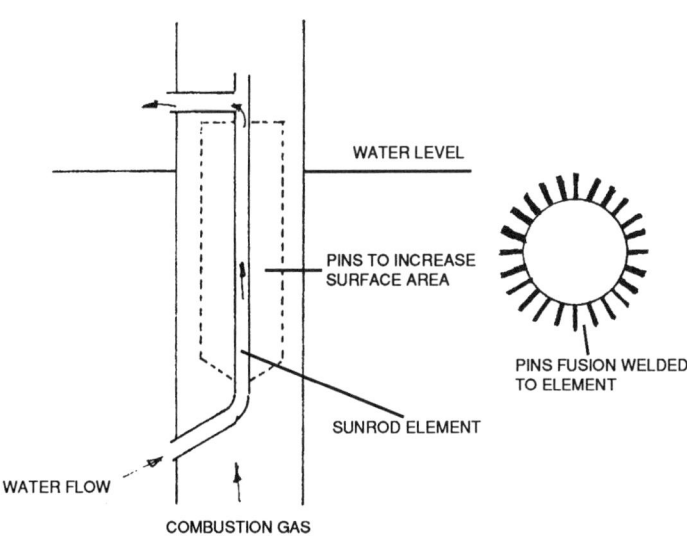

b) SUNROD ELEMENT

is also relatively easy and is accomplished, when the boiler is shut down, by simply removing the cleaning doors, opening the drain and spraying with high pressure fresh water.

Pressure control of the steam is accomplished by flashing the boiler when pressure drops below a pre-set level during periods of high steam load and, dumping steam to the condenser when the pressure rises due to low steam load.

Vapour Vertical Boiler (Coiled-Tube)

Fig. 147. shows in a simplified diagrammatic form a coiled-tube boiler of the Stone-Vapor type. It is compact, space saving, designed for u.m.s. operation, and is supplied ready for connecting to the ships services.

A power supply, depicted here by a motor, is required for the feed pump, fuel pump (if fitted), fan and controls.

Feed water is force circulated through the generation coil wherein about 90% is evaporated. The un-evaporated water travelling at high velocity carries sludge and scale into the separator, which can be blown out at intervals manually or automatically. Steam at about 99% dry is taken from the separator for shipboard use.

The boiler is completely automatic in operation. If, for example, the steam demand is increased, the pressure drop in the separator is

FIG 147
PACKAGE COIL TYPE BOILER

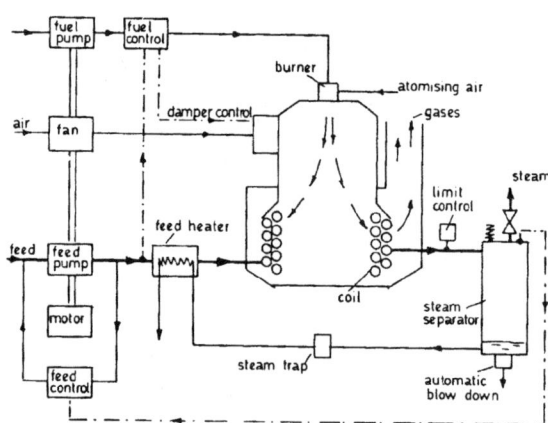

sensed and a signal, transmitted to the feed controller, demands increased feed, which in turn increases air and fuel supply. With such a small water content explosion due to coil failure is virtually impossible and a steam temperature limit control protects the coil against abnormally high temperatures. In addition the servo-fuel control protects the boiler in the event of failure of water supply. Performance of a typical unit could be:
- Steam pressure 10 bar
- Evaporation 3000 kg/h
- Thermal efficiency 80%
- Full steam output in about 3 to 4 mins.

Note: Atomising air for the fuel may be required at a pressure of about 5 bar.

Steam to Steam Generation
In vessels which are fitted with water tube boilers a protection system of steam to steam generation may be used instead of desuperheaters and reducing valves, etc. (See later.)

TURBO GENERATORS

Such turbines are fairly standard l.p. steam practice and reference, where necessary, could be made to Volume 9. Detailed instructions are provided on board ship for personnel unfamiliar with turbine practice. For the purposes of this chapter the short extract description given below should be typical and adequate.

Turbine
A single cylinder, single axial flow, multistage (say 5) impulse turbine provided with steam through nozzles at 10 bar and 300°C preferably with superheat to limit exhaust moisture to 12%. Axial adjustment of rotor position is usually arranged at the thrust block and protection for overspeed, low oil pressure and low vacuum are provided. Materials and construction for the turbine unit and single reduction gearing are standard modern practice.

Electrical
The turbine at 100-166 rev/s drives the alternator and exciter through a reduction of about 6:1 to produce typically 450-600 kW at 440 V, 3 ph., 60 Hz. A centrifugal shaft-driven motorised

governor arranged for local or switchboard operation would operate the throttle valve via a hydraulic servo. Straight line electrical characteristics normally incorporates a speed droop adjustment to allow ready load sharing with auxiliary diesel generators or an extra turbo unit.

Ancillary Plant

This is normally provided as a package unit with condenser, air ejector, auto gland seals, gland condenser, motorised and worm-driven oil pumps, etc. A feed system is provided either integral or divorced from the turbine-gearbox-alternator unit. Exhaust can be arranged to a combined condenser incorporating cargo exhaust. Control utilises gas by-pass, dumping steam, etc.

SILENCERS

Normally waste heat boilers act as spark arresters and silencers at all times. The silencer sketched in Fig. 148. would not usually be fitted if such boilers were used but a short description of the silencer may be useful.

Three designs have been utilised. The tank type has a reservoir of volume about 30 times cylinder volume. Baffles are arranged to give about four gas reversals. The diffuser type has a central perforated discharge pipe surrounded by a number of chambers of varying volume. The orifice type is sketched in Fig. 148. and the construction should be clear. Energy pulsations and sound waves are dissipated by repeated throttling and expansion.

GAS ANALYSIS

A number of factors have been stated which affect the design and operation of the plant and some salient points will now be briefly considered.

Optimum Pressure

This depends on the system adopted but in general the range is from 6 bar to 11 bar. The lower pressures give a cheaper unit with near maximum heat recovery. However higher pressures allow more flexibility in supply with perhaps more useful steam for

FIG 148
SILENCER AND SPARK ARRESTER

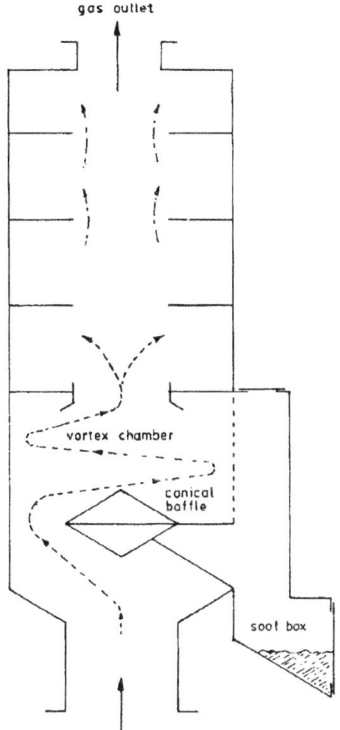

certain auxiliary functions together with reserve steam capacity to meet variations in demand. Low feed inlet temperatures reduce pressure and evaporative rate.

Temperature

A minimum temperature differential obviously applies for heat transfer. Temperature difference, fouling, gas velocity, gas distribution, metal surface resistance, etc., are all important factors. Reduction in service engine revolutions will cause reduced gas mass and temperature increase if the power is maintained constant. A similar effect will be apparent under operation in tropical conditions. The effect of increased back pressure will be to raise the gas temperature for a given air inlet temperature. Fig. 149. illustrates: (a) typical heat transfer diagram, and (b) gas temperature/mass-power curves. A common temperature differential is about 40°C, *i.e.* water inlet 120°C and gas exit 160°C.

FIG 149
EXHAUST GAS CONDITIONS

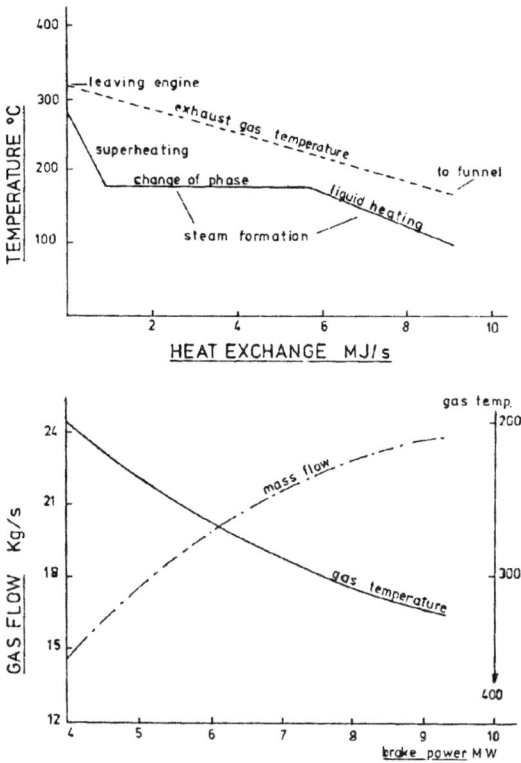

WASTE HEAT PLANT

Corrosion
The acid dew point expected is about 110°C with a 3% sulphur fuel and a high rate of conversion from SO_2 to SO_3 is possible. Minimum metal temperatures of 120°C for mild steel are required.

Exhaust System
The arrangement must offer unrestricted flow for gases so that back pressure is not increased. Good access is required for inspection and cleaning. On designs with alternate gas-oil firing provision must be made for quick and foolproof change-over with no possibility of closure to atmosphere and waste heat system at the same time.

GAS/WATER HEAT EXCHANGERS

Waste Heat Economisers
Such units are well proven in steamship practice and similar all-welded units are reliable and have low maintenance costs in motorships. Gas path can be staggered or straight through with extended surface element construction. Large flat casings usually require good stiffening against vibration. Water wash and soot blowing fittings may be provided.

Waste Heat Boilers
These boilers have a simple construction and fairly low cost. At this stage a single natural circulation boiler will be considered and these normally classify into three types, namely: simple, alternate and composite.

Simple
These boilers are not very common as they operate on waste heat only. Single or two-pass types are available, the latter being the most efficient. Small units of this type have been fitted to auxiliary oil engine exhaust systems, operating mainly as economisers, in conjunction with another boiler. A gas change valve to direct flow to the boiler or atmosphere is usually fitted as described below.

Alternate
This type is a compromise between the other two. It is arranged to give alternate gas and oil firing with either single or double pass

gas flow. It is particularly important to arrange the piping system so that oil fuel firing is prevented when exhaust gas is passing through the boiler. A large butterfly type of change-over valve is fitted before the boiler so as to direct exhaust gas to the boiler or to the atmosphere. The valve is so arranged that gas flow will not be obstructed in that as the valve is closing one outlet the other outlet is being opened. The operating mechanism, usually a large external square thread, should be arranged so that with the valve directed to the boiler, fuel oil is shut off. A mechanical system using an extension piece can be arranged to push a fork lever into the operating handwheel of the oil fuel supply valve. When the exhaust valve is fully operated to direct the gas to atmosphere the fork lever then clears the oil fuel valve handwheel after change-over travel is completed. It is also very important to ensure full fan venting and proper fuel heating-circulation procedures before lighting the oil fuel burners.

Composite

Such boilers are arranged for simultaneous operation on waste heat and oil fuel. The oil fuel section is usually only single pass. Early designs utilised Scotch boilers, with, say, a three furnace boiler, it may mean retaining the centre or the wing furnaces for oil fuel firing. The gas unit would often have a lower tube bank in place of the furnace, with access to the chamber from the boiler back, so giving double pass. Alternative single pass could be arranged with gas entry at the boiler back. Exhaust and oil fuel sections would have separate uptakes and an inlet change-over valve was required. In general Scotch boilers as described are nearly obsolete and vertical boilers are used. As good representative, and more up-to-date, common practice, two types of such boiler will be considered.

Cochran Boiler

The Cochran boiler whose working pressure is normally of the order of 8 bar is available in various types and arrangements, some of which are:

Single pass composite, *i.e.* one pass for the exhaust gases and two uptakes, one for the oil fired system and one for exhaust system. Double pass composite, *i.e.* two passes for the exhaust

FIG 150
DIAGRAMMATIC ARRANGEMENT OF A SINGLE PASS COMPOSITE COCHRAN BOILER

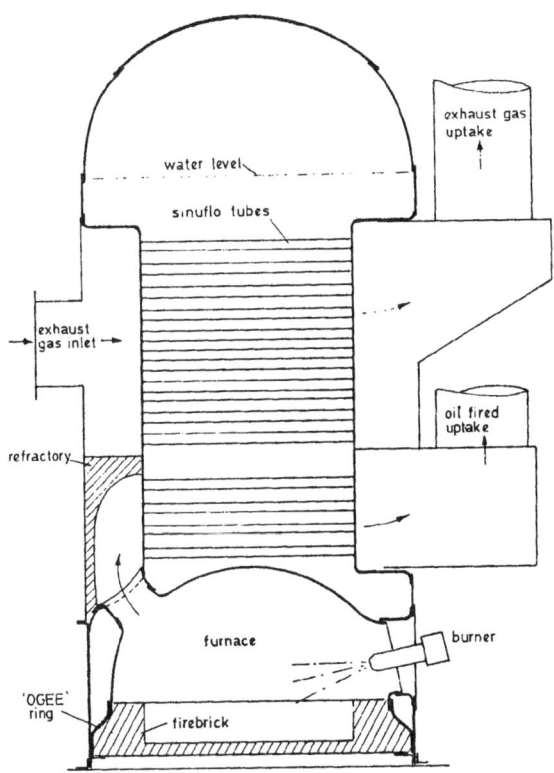

gases and two uptakes, one for the oil fired system and one for the exhaust system. (Double pass exhaust gas, no oil fired furnace and a single uptake, is available as a simple type. Or, double pass alternatively fired, *i.e.* two passes from the furnace for either exhaust gases or oil fired system with one common uptake).

The boiler is made from good quality low carbon open hearth mild steel plate. The furnace is pressed out of a single plate and is therefore seamless.

Connecting the bottom of the furnace to the boiler shell plating is a seamless 'Ogee' ring. This ring is pressed out of thicker plating than the furnace, the greater thickness is necessary since circulation in its vicinity is not as good as elsewhere in the boiler and deposits can accumulate between it and the boiler shell plating. Hand hole cleaning doors are provided around the circumference of the boiler in the region of the 'Ogee' ring.

The tube plates are supported by means of tube stays and by gusset stays, the gusset stays supporting the flat top of the tube plating.

Tubes fitted, are usually of special design (Sinuflo), being smoothly sinuous in order to increase heat transfer by promoting turbulence. The wave formation of the tubes lies in a horizontal plane when the tubes are fitted, this ensures that no troughs are available for the collection of dirt or moisture. This wave formation does not in any way affect cleaning or fitting of the tubes.

Thimble Tube Boiler

There are various designs of thimble tube boiler, these include: oil fired, exhaust gas, alternatively fired and composite types.

The basic principle with which the thimble tube operates was discovered by Thomas Clarkson. He found that a horizontally arranged tapered thimble tube, when heated externally, could cause rapid ebullitions of a spasmodic nature to occur to water within the tube, with subsequent steam generation. Fig. 151 shows diagrammatically an alternatively fired boiler of the Clarkson thimble tube type capable of generating steam with a working pressure of 8 bar. The cylindrical outer shell encloses a cylindrical combustion chamber, from which, radially arranged thimble tubes project inwards. The combustion chamber is attached to the bottom of the shell by an 'Ogee' ring and to the top of the shell by a cylindrical uptake. Centrally arranged in the combustion chamber is an adjustable gas baffle tube.

EXHAUST GAS HEAT RECOVERY CIRCUITS

Many circuits are possible and a few arrangements will now be considered. Single boiler units as discussed, whilst cheap, are not flexible and have relatively small steam generating capacity. The systems now considered are based on multi-boiler installations.

FIG 151
ALTERNATIVELY FIRED THIMBLE-TUBE BOILER

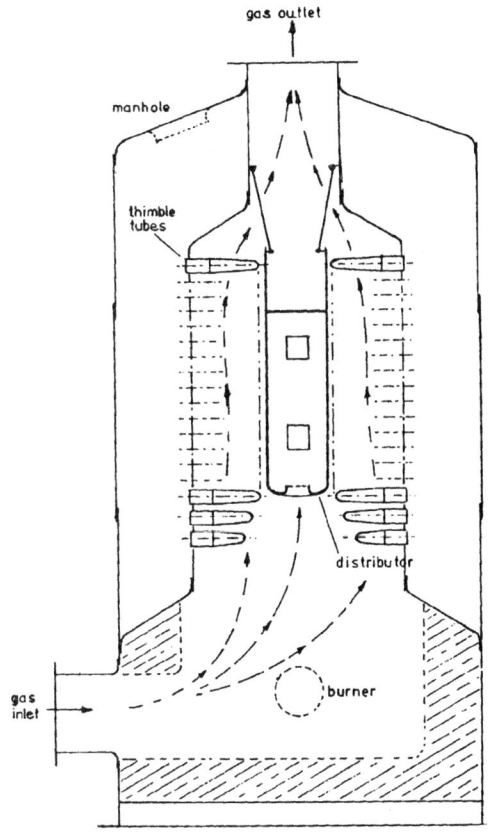

Natural Circulation Multi-Boiler System
It is possible to have a single exhaust gas boiler located high up in the funnel, operating on natural circulation whereby a limited amount of steam is available for power supply whilst the vessel is at sea. In port or during excessive load conditions, the main boiler or boilers are brought into operation to supply steam to the same steam range by suitable cross connecting steam stop valves. In port

FIG 152
NATURAL CIRCULATION/WASTE HEAT PLANT AND W.T. BOILER

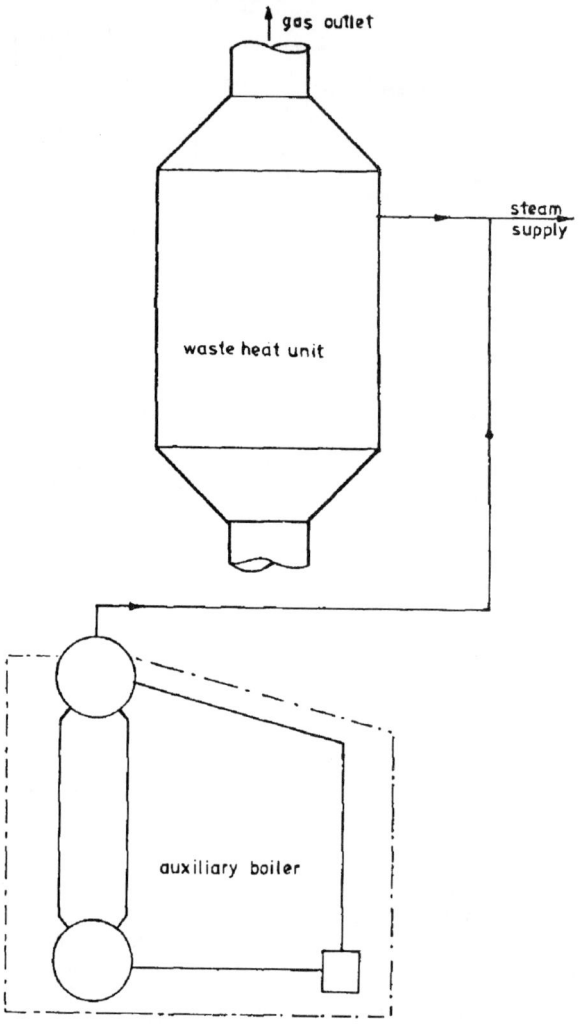

FIG 153
NATURAL CIRCULATION

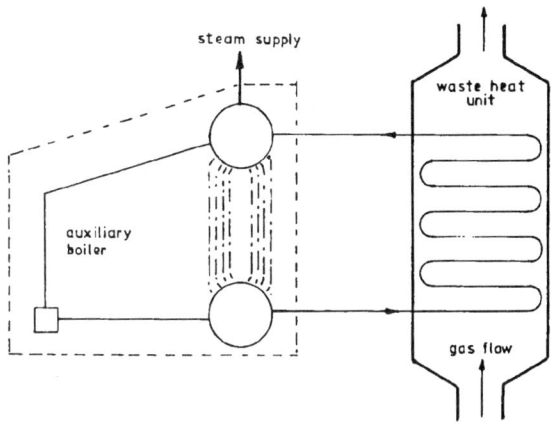

the exhaust gas boiler is secured and all steam supplied by the oil fired main boilers. This system is suitable for use on vessels such as tankers where a comparatively large port steaming capacity may be required for operation of cargo pumps, but suffers from the disadvantage that the main boilers must either be warmed through at regular intervals or must be warmed through prior to reaching port. Further to this the main boilers are not immediately ready for use in event of an emergency stop at sea unless the continuous warming through procedure has been followed.

Forced Circulation Multi-Boiler System

In order to improve the heat transfer efficiency and to overcome the shortcomings of the previous example a simple forced circulation system may be employed. The exhaust gas boiler is arranged to be a drowned heat exchanger which, due to the action of a circulating pump, discharges its steam and water emulsion to the steam drum of a water-tube boiler. The forced circulation pump draws from near the bottom of the main boiler water drum and circulates water at almost ten times the steam production rate so giving good heat transfer. The steam/water emulsion on being discharged into the water space of the main boiler drum separates out exactly in the same way as if the boiler were being oil fired. This arrangement ensures that the main boiler is always warm and

FIG 154
FORCED CIRCULATION WASTE HEAT PLANT
AND MAIN BOILER

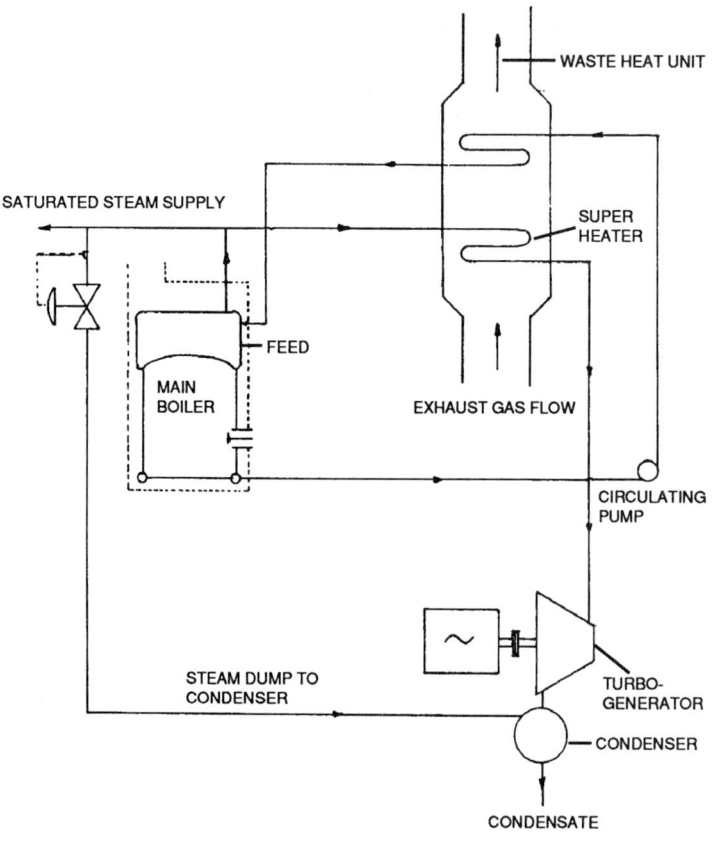

capable of being immediately fired by manual operation or supplementary pilot operated automatic fuel burning equipment. Feed passes to the main boiler and becomes neutralised by chemical water treatment. Surface scaling is thus largely precluded and settled out impurities can be removed at the main boiler blowdown. If feed flow only is passed through an economiser type unit parallel flow reduces risks of vapour locking. Unsteady feed flow at normal gas conditions can result in water flash over to steam and rapid metal temperature variations. Steam, hot water and cold water conditions can cause thermal shock and water hammer.

Contra flow designs are generally more efficient from a heat transfer viewpoint giving gas temperatures nearer steam temperature and are certainly preferred for economisers if circulation rate is a multiple of feed flow. The generation section is normally parallel flow and the superheat section contra flow. Output control could be arranged by output valves at two different levels so varying the effective heat transfer surface utilised. In addition a circulating pump by-pass arrangement gives an effective control method.

Dual Pressure Forced Circulation Multi-Boiler System

This concept has been incorporated in the latest waste heat circuits and the sketch illustrates how the general principle can be applied in conjunction with a waste heat exchanger to supply superheated steam. By this means every precaution has been taken to minimise the effect of contamination of the water-tube boiler.

Steam generated in the water-tube boiler by either oil firing or waste heat exchanger passes through a submerged tube nest in the steam/steam generator to give lower grade steam which is subsequently passed to the superheater.

A water-tube boiler, steam/steam generator and feed heater may be designed as a packaged unit with the feed heater incorporated in the steam/steam generator. The high pressure high temperature system at say 10 bar will supply a turbo generator for all electrical services while the low pressure system at say $2^{1/2}$ bar would provide all heating services. Obviously the dual system is more costly. Numerous designs are possible including separate lp and hp boilers, either natural or forced circulation, indirect systems with single or double feed heating, etc.

FIG 155
DUAL PRESSURE SYSTEM

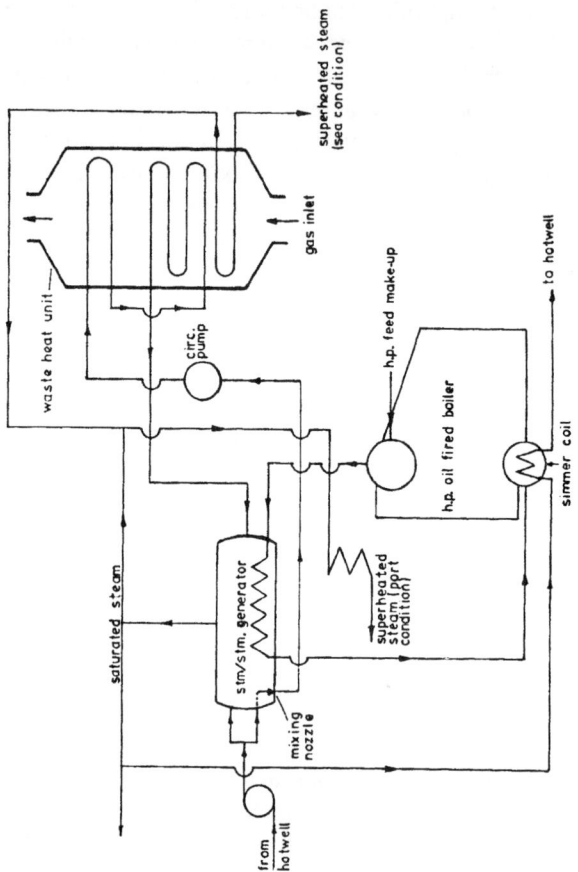

WATER/WATER HEAT EXCHANGERS

Evaporators

The basic information given on evaporators in Volume 8 should *first* be considered. In motor ship practice efficient single effect units incorporating flexible elements and controlled water level are in service. Evaporators utilising jacket cooling water as the heating medium producing an output of 20-25 tonnes per day are common. Flash evaporators have increasingly been fitted on large vessels utilising multi-stage units. Also multiple effect evaporators of conventional form are used. The steam circuit of many modern motor ships has developed in complexity to approach successful steam ship practice.

FEED HEATING

The advantages of pre-heating feed water are obvious. Three methods will be considered, namely: economisers, mixture and indirect. Economiser types have been included in previous discussion and sketches. It is sufficient to repeat that such systems require a careful design to cope with fluctuations of steam demand and that particular attention is necessary to ensure protection against corrosive attack. Mixture systems employ parallel feeding with circulating pump and feed pump to the economiser inlet. Such circuits require careful matching of the two pumps and control has to be very effective to prevent cold water surges leading to reducing metal temperatures and causing corrosion. Indirect systems require a water/water exchanger feed heater.

This design reduces the risk of solid deposit in the economiser and maintains steady conditions of economiser water flow so protecting the economiser against corrosive attack. A typical system is shown in Fig. 156. If boiler pressure tends to rise too high the circulation by-pass will be opened.

The effect will be twofold, *i.e.* feed water will enter the boiler at the lower temperature and water temperature entering the economiser is at a higher temperature. These two effects serve to reduce boiler pressure and so control the system. Obviously this system is more costly but is very flexible.

COMBINED HEAT RECOVERY CIRCUITS

The low grade heat of engine coolant systems restricts the heat recovery in such secondary circuits to temperatures near 7-8°C. As such it is normally restricted to use with distillation plants. Combined or compound units involving combination between engine coolant and exhaust gas systems are complicated by the need to prevent contamination and utilise the large volume of low temperature coolant in circulation. Jacket water coolant temperatures have increased in recent motorship practice but even if the engine design can be modified to suit even higher temperatures there is always a problem of high radiation heat loss from jackets to confined engine rooms.

**FIG 156
INDIRECT FEED HEATING**

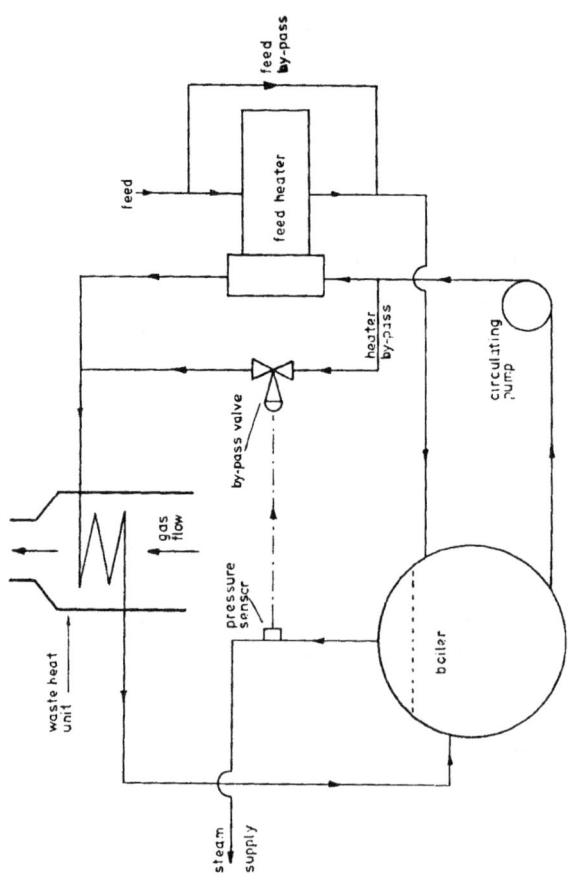

CHAPTER 10

MISCELLANEOUS

CRANKCASE EXPLOSIONS

Introduction

The student should first refer to Volume 8 for a consideration of spontaneous ignition temperatures and also limits of inflammability in air. Crankcase explosions have occurred steadily over the years with perhaps that of the *Reino-del-Pacifico* in 1947 the most serious of all. In fact crankcase explosions have occurred in all types of enclosed crankcase engines, including steam engines. Explosions occur in both trunk piston types and in types with a scraper gland seal on the piston rod. Much research has been done in this field but the difficulties of full experimentation utilising actual engines under normal operating conditions is almost impossible to attempt. The following is a simplified presentation based on the mechanics of cause of explosion, appropriate DoT regulations and recommendations and descriptive details of preventative and protective devices utilised.

Mechanics of Explosion

1. A hot spot is an essential source of such explosions in crankcases as it provides the necessary ignition temperature, heat for oil vapourisation and possibly ignition spark. Normal crankcase oil spray particles are in general too large to be easily explosive (average 200 microns). Vapourised lubricating oil from the hot source occurs at 400°C, in some cases lower, with a particle size explosive with the correct air ratio (average 6 microns). Vapour can condense on colder regions, a condensed mist with fine particle size readily causes explosion in the presence of an ignition source. A lower limit of flammability of about 50 mg/l is often found in practice. Experiment indicates two separate temperature regions in which ignition can take place, *i.e.* 270°C-350°C and above 400°C.

2. Initial flame speed after mist ignition is about 0·3 m/s but unless the associated pressure is relieved this will increase to about

300 m/s with corresponding pressure rise. In a long crankcase, flame speeds of 3 km/s are possible giving detonation and maximum damage. The pressure rise varies with conditions but without detonation does not normally exceed 7 bar and may often be in the range 1-3 bar.

3. A primary explosion occurs and the resulting damage may allow air into a partial vacuum. A secondary explosion can now take place, which is often more violent than the first followed by similar sequence until equilibrium.

4. The pressure generated, as considered over a short but finite time, is not too great but instantaneously is very high. The associated flame is also dangerous. The gas path cannot ordinarily be deflected quickly due to the high momentum and energy.

5. Devices of protection must allow gradual gas path deflection, give instant relief followed by non return action to prevent air inflow and be arranged to contain flame and direct products away from personnel.

6. Delayed ignition is sometimes possible. An engine when running with a hot spot may heat up through the low temperature ignition region without producing flame because of the length of ignition delay period at low temperatures. Vapourised mist can therefore be present at 350-400°C. If the engine is stopped the cooling may induce a dangerous state and explosion. Likewise air ingress may dilute a previously too rich mixture into one of dangerous potential.

7. Direct detection of overheating by thermometry offers the greatest protection but the difficulties of complete surveillance of all parts is prohibitive.

8. A properly designed crankcase inspection door preferably bolted in place, suitably dished and curved with say a 3 mm thickness of sheet steel construction should withstand static pressures up to 12 bar although distorted.

9. There are many arguments for and against vapour extraction by exhauster fans. There is no access of free air to the crankcase and the fan tends to produce a slight vacuum in the crankcase. On balance most opinion is that the use of such fans can reduce risk of explosion. The danger of fresh air drawn into an existing over rich heated state is obvious. On the practical aspect leakage of oil is reduced.

Crankcase Safety Arrangements

The following are based on specific DoT rules:

1. Means should be adapted to prevent danger from the result of explosion.

2. Crankcases and inspection doors should be of robust construction. Attachment of the doors to the crankcase (or entablature) should be substantial.

3. One or more non-return pressure relief valves should be fitted to the crankcase of each cylinder and to any associated gearcase.

4. Such valves should be arranged or their outlets so guarded that personnel are protected from flame discharge with the explosion.

5. The total clear area through the relief valves should not normally be less than 9·13 cm^2/m^3 of gross crankcase volume.

6. Engines not exceeding 300 mm cylinder bore with strongly constructed crankcases and doors may have relief valve or valves at the ends only. Similarly constructed engines not exceeding 150 mm cylinder bore need not be fitted with relief valves.

7. Lubricating oil drain pipes from engine sump to drain tank should extend to well below the working oil level in the tank.

8. Drain or vent pipes in multiple engine installations are to be so arranged so that the flame of an explosion cannot pass from one engine to another.

9. In large engines having more than six cylinders it is recommended that a diaphragm should be fitted at near mid length to prevent the passage of flame.

10. Consideration should be given to means of detection of over-heating and injection of inert gas.

The above should be self explanatory in view of the previous comments made.

Preventative and Protection Devices

In general three aspects are worthy of consideration, *i.e.* relief of explosion, flame protection and explosive mist detection.

Crankcase Explosion Door

A design is shown in Fig. 157. The sketch illustrates a combined valve and flame trap unit with the inspection door insertion in the middle. The internal section supports the steel gauze element and the spider guide and retains the spindle. The external combined aluminium valve and deflector has a synthetic rubber seal. Pressure setting on such doors is often 1/15 bar (above atmospheric pressure). Relief area and allowable pressure rise vary with the licensing insurance authority but a metric ratio of 1:90 should not normally be exceeded based on gross crankcase volume and this should not allow explosion pressures to exceed about 3 bar.

Flame trap
Such devices are advisable to protect personnel. The vented gases can quickly be reduced in temperature by gauze flame traps from say 1500°C to 250°C in 0·5 m. Coating on the gauzes, greases or engine lubricating oil, greatly increases their effectiveness. The best location of the trap is inside the relief valve when it gives a more even distribution of gas flow across its area and liberal wetting with lubricating oil is easier to arrange. A separate oil supply for this action may be necessary. The explosion door in Fig. 157. has an internal mesh flame trap fitted.

Flame traps effectively reduce the explosion pressure and prevent two stage combustion. Gas-vapour release by the operation of an oil wetted flame trap is not usually ignitable. Typical gauze mild steel wire size is 0·3 mm with 40% excess clear area over the valve area.

Crankcase Oil Mist Detector

If condensed oil mists are the sole explosive medium then photo-electric detection should give complete protection but if the

FIG 157
CRANKCASE EXPLOSION RELIEF DOOR

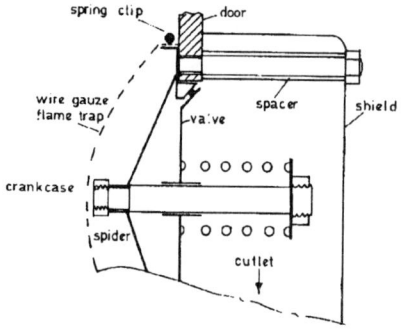

crankcase spray is explosive the mist detection will only indicate a potential source of ignition. The working of one design of detector should be fairly clear from Fig. 158. The photo cells are normally in a state of electric balance, *i.e.* measure and reference tube mist content in equilibrium. Out of balance current due to rise of crankcase mist density can be arranged to indicate on a galvanometer which can be connected to continuous chart recording and auto visual or audible alarms. The suction fan draws a large volume of slow moving oil-air vapour mixture in turn from various crankcase selection points. Oil mist near the lower critical density region has a very high optical density. Alarm is normally arranged to operate at $2^{1/2}\%$ of the lower critical point, *i.e.* assuming 50 mg/l as lower explosive limit then warning at 1·25 mg/l.

Operation
The fan draws a sample of oil mist through the rotary valve from each crankcase sampling pipe in turn, then though the measuring tube and delivers it to atmosphere. An average sample is drawn from the rotary valve chamber through the reference tube and delivered to atmosphere at the same time. In the event of overheating in any part of the crankcase there will be a difference in optical density in the two tubes, hence less light will fall on the photo cell in the measuring tube. The photo cell outputs will be different and when the current difference reaches a pre-determined value an alarm signal is operated and the slow turning rotary valve

FIG 158
CRANKCASE OIL MIST DETECTOR

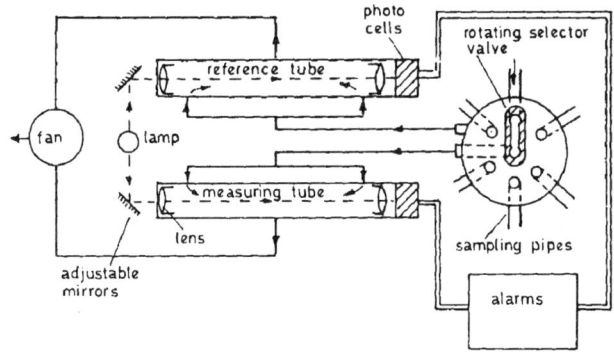

stops, indicating the location of the overheating.

Normal oil particles as spray are precipitated in the sampling tubes and drain back into the crankcase.

CO_2 Drenching System

30% by volume of this inert gas is a complete protection against crankcase explosion. This is particularly beneficial during the dangerous cooling period. Automatic injection can be arranged at, say 5% of critical lower mist density but in practice many engineers prefer manual operation. When the engine is opened up for inspection, and repair at hot source, it will of course be necessary to ensure proper venting before working personnel enter the crankcase.

EXHAUST GAS POWER TURBINE

In an effort to improve overall plant efficiency the turbocharger manufacturer ABB has developed a system which exploits surplus exhaust gas in a power turbine and is fed either, to the engine crankshaft or, to an auxiliary diesel or steam turbine generator. Fig. 159. The latter is only feasible if the demand for electricity is greater than the output of the power turbine. This development has been made possible by the improved turbocharger efficiencies achieved in recent years resulting in surplus energy being available in the exhaust gas.

The power turbine can be brought in and out of service, as conditions require, by operating a flap in the exhaust line.

The turbine part of the power turbine is similar to those of turbochargers. The drive from the turbine is via epicyclic gearing and a clutch to the chosen mode of power input.

GAS TURBINES

The gas turbine theoretical cycle and simple circuit diagram have been considered in Chapter 1. Marine development of gas turbines stemmed from the aero industry in the 1940's. Apart from an early stage of rapid progress the application to marine use has been relatively slow until recently. Consideration can best be applied in two sections, namely, industrial gas turbines and aero-derived types. In general this can largely be considered as a 'marinisation' of equipment originally designed for other duty.

FIG 159
RECOVERY OF SURPLUS EXHAUST GAS ENERGY IN A POWER TURBINE

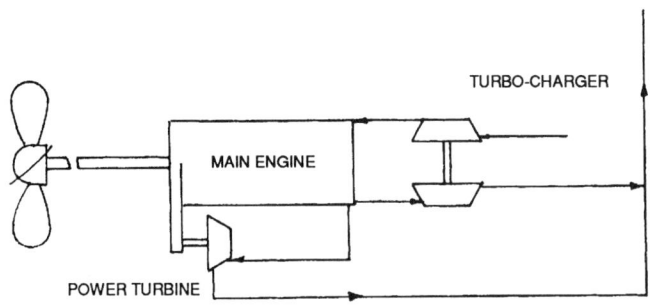

POWER TURBINE COUPLED TO MAIN ENGINE

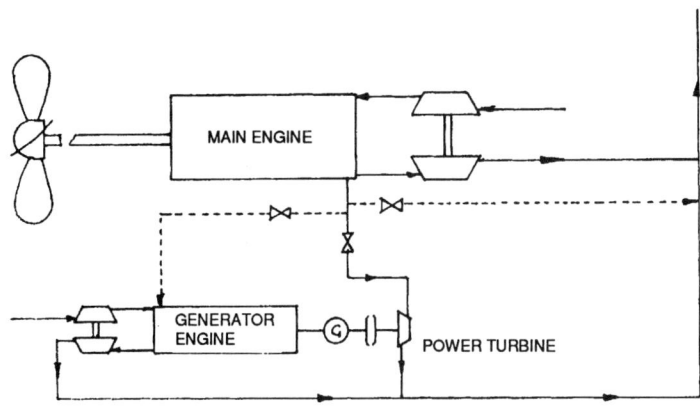

POWER TURBINE COUPLED TO A GENERATOR ENGINE

Industrial Gas Turbines
The simplest design is a single-shaft unit which has low volume and light weight (5 kg/kW at 20,000 kW). Fuel consumption (specific) may be about 0·36 kg/kWh on residual fuels. This consumption is not normally acceptable for direct propulsion and initial usage was as emergency generators in MN practice and the RN for small vessels or as boost units in larger warships. Compared to steam turbines (32% output, 58% condenser loss) the simple gas turbine (24% output, 73% exhaust loss) is less efficient but the addition of exhaust gas regeneration gives 31% output (specific fuel consumption 0·28 kg/kWh) and combined RN units 36% output. Normally a two-shaft arrangement was preferred in MN practice in which load shaft and compressor shaft are independent.

A design was available by 1955 for main propulsion with maximum turbine inlet conditions of 6 bar, 650°C and specific fuel consumption approaching 0·3 kg/kWh. Starting of the twin-shaft unit was by electric motor, power variation by control of gas flow, conventional gear reduction and propeller drive by hydraulic clutch with astern torque converter (more modern practice uses variable pitch propeller). Turbine and turbo-compressor design utilised standard theory and simple module construction utilising horizontally split casings, diffusers, etc., and easily accessible nozzles.

To improve efficiency even further it is necessary to use much higher inlet gas temperatures (1200°C would give a specific fuel consumption of about 0·2 kg/kWh). The limiting factor is suitable materials. Experiments have been, and still are, being carried out with ceramic blades and with cooled metallic blades. Essentially the problem is the same for steam turbine plant and there has been no marked incentive for the shipowner to install gas turbine plant in preference to equally economic and established steam systems. During the 1960's experience was established in the vessels *Auris, John Sergeant* and *William Paterson.* It may well be that direct gas cooled reactors in conjunction with closed cycle gas turbines in electric power generation may be an attractive possibility in Nuclear technology. G.E.C. produce a wide range (4000-50,000 kW) of industrial gas turbines now effectively marinised for marine propulsion. In addition to reliability, easy maintenance, low

FIG 160
OPEN CYCLE MARINE GAS TURBINE SYSTEM

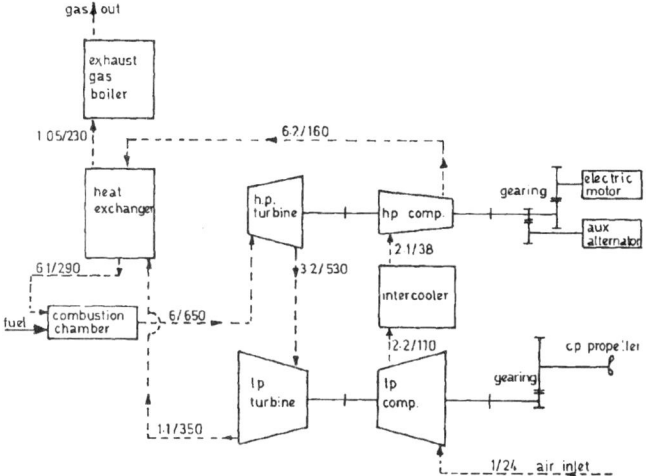

volume, etc., the very easy application to electric drives and to automation make the units attractive. Geared drive usually utilises locked train helical gears or alternatively epicyclic gearing.

C.P. propeller development has also broadened the possibilities of various propulsion systems, including geared diesel – gas turbine systems. Marine gas turbines do run with a high noise level and they require to be water washed at regular intervals, the latter depending upon the type of fuel being used. The recently changing design of ships has meant that the owner, or operator, needs to analyse propulsion systems carefully for all economic factors, which vary greatly for VLCC, Ro-Ro, LNG, container vessels, etc. Gas turbines have been exclusively adopted for RN surface vessels.

Aero-Derived
Apart from RN units so derived from aero gas turbines the first British MN vessel so engined was the g.t.s. *Euroliner* in 1970. Turbo Power and Marine Systems Inc. twin gas turbines, 22, 500 kW each at 3600 rev/min drive separate screw shafts at 135

FIG 161
NITROGEN CHEMICAL ACTIVITY INCREASING WITH TEMPERATURE

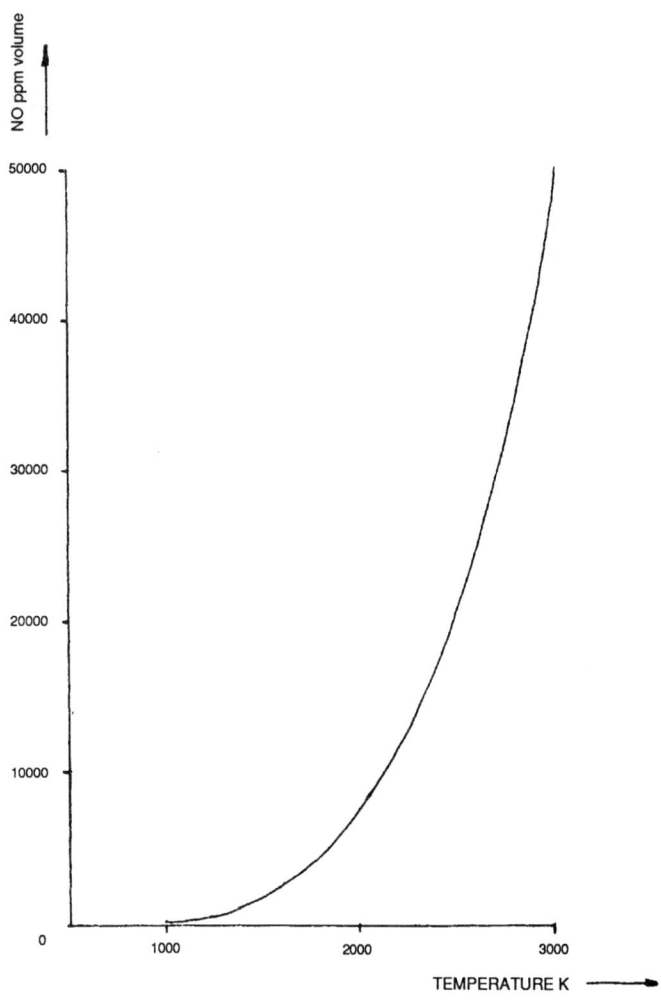

FIG 162
SELECTIVE CATALYTIC REDUCTION (SCR) APPARATUS

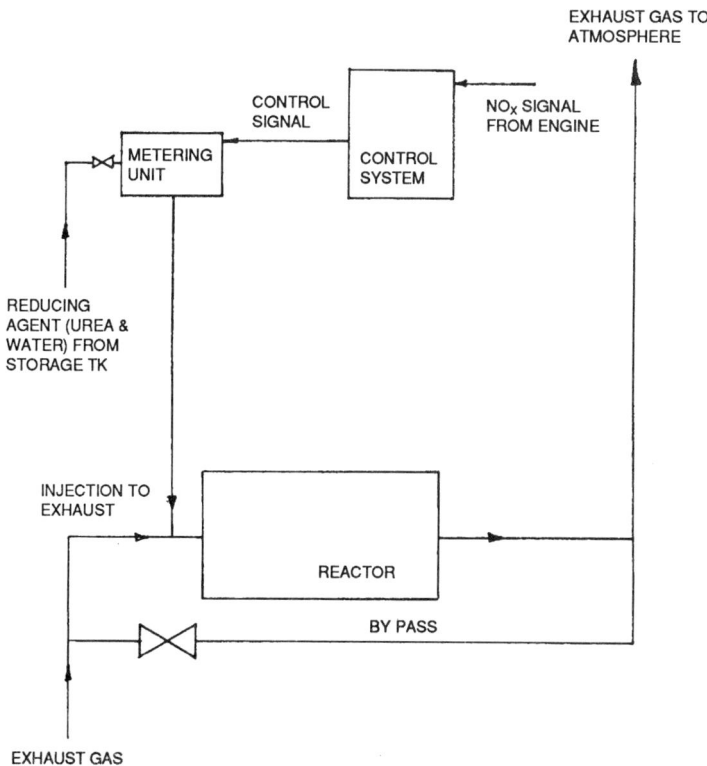

rev/min through double reduction locked train gears, with controllable pitch propellers. Main electrical alternators are driven from the gearbox.

Exhaust Gas Emissions

Small quantities of toxic substances such as oxides of sulphur and nitrogen occur in diesel engine exhaust gas. The oxides of sulphur, often referred to as SO_x are formed in quantities proportional to the sulphur content of the fuel. It is likely that the only practical method of reducing the level of SO_x in the exhaust is by consuming fuel of low sulphur content. The oxides of nitrogen, often referred to as NO_x, are the most significant emission from marine diesel engines and comprise nitrous oxide [NO] and nitrogen dioxide [NO_2]. Loaded diesels produce very low amounts of NO_2 directly, over 90% of NO_x being NO. Although the NO produced by diesel engines is considered harmless it oxidises in the atmosphere to form the more environmentally harmful NO_2. The NO_x are formed from two sources:

1. When nitrogen in the fuel combines with the oxygen in the combustion air [possibly 20% of the total NO_x present]

2. Nitrogen and oxygen in the combustion air.

In previous studies students may have considered nitrogen as being inert and not taking part in the combustion process. This, however, is a simplification and when dealing with exhaust emission, nitrogen cannot be thought of as inert. Studies have shown that the chemical activity of nitrogen increases with temperature. Fig. 161. Modern diesels achieve their high thermal efficiency as a result of burning the fuel at high temperatures. Thus, the more efficient the engine the greater the amount of NO_x the engine is likely to generate.

Low NO_x generation can be achieved:

1. By ensuring that the fuel/air mixture is as homogeneous as possible in order to keep the percentage of rich mixture in the combustion chamber small. This requires a nozzle with as many spray holes as possible.

2. By using a high air/fuel ratio.

3. By retarding the fuel injection.

Only the first of these options does not involve a fuel penalty. Raising the air/fuel ratio beyond what is necessary leads to increased pumping losses. Retarding the injection results in lower firing pressures. These measures, however, have the advantage that they do not require extra components to be added to the engine.

Other measures which do require extra components and added cost include.

1. Adding water to the fuel. This works in two ways, it helps to create a homogeneous fuel/air mixture and also reduces the temperature of combustion. The injection of water requires an increase of 20-30% in fuel pump capacity. Precautions must also be taken to maintain the oil/water emulsion in a stable condition and to prevent corrosion of the fuel system components.

2. The use of Selective Catalytic Reduction [SCR].

Selective Catalytic Reduction

SCR technology was developed for land based installations and is being developed for main applications and involves injecting small amounts of a single atom nitrogen based additive, such as ammonia [NH_3], into the exhaust. Because of the difficulties and dangers of handling ammonia on board ship a safer more easily handled ammonia compound called urea {$2[NH_2]CO$} is being used. The principle is to combine the nitrogen atoms of the NO_x and NH_3 compound to form a stable nitrogen [N_2] molecule which is the main constituent of air.

The principle catalytic reduction process of the ammonia compound is according to the following chemical reaction.

$$2NO + 2NH_3 + 1/2O_2 + CO_2 \xrightarrow{catalysis} 2N_2 + 3H_2O + CO_2$$

To accomplish this at the temperatures encountered in exhaust gas – 250-450°C a catalyst is required. The catalyst is an oxide of vanadium carried on a heat resistant honeycomb of ceramic. In order to minimise the pressure drop across the reactor the gas passages of the honeycomb core must be of sufficient cross sectional area.

The urea is mixed with water and metered into the exhaust gas upstream of the reactor at a rate dependent upon engine load. Fig. 162.

TEST QUESTIONS

(S denotes SCOTVEC questions)

CHAPTER 1– CLASS ONE

1. *(a)* With reference to fatigue of engineering components explain the influence of stress level and cyclical frequency on expected operating life.
   (b) Explain the influence of material defects on the safe operating life of an engineering component. S
   (c) State the factors which influence the possibility of fatigue cracking of a bedplate transverse girder and explain how the risk of such cracking can be minimised.

2. With reference to engine performance monitoring discuss the relative merits of electronic indicating equipment when compared with traditional indicating equipment.

3. As Chief Engineer Officer how would you ascertain if the main engine is operating in an overloaded condition?
   If the engine is overloading what steps would you take to ensure that the engine was brought within the correct operating range?

4. A set of indicator cards suggests that individual cylinder powers of the main engine are not balanced:
   What action would you take to rectify the problem?
   How would you ascertain the accuracy of the cards?

5. Explain, by referring to the theoretical considerations how the efficiency of an IC engine is dependent upon the compression ratio.
   Why does an actual engine power card only approximate to the ideal cycle?

CHAPTER 2– CLASS ONE

1. *(a)* State with reasons the main causes of normal and abnormal cylinder liner wear.
   (b) State the ideal properties of a cylinder oil for use with an engine burning residual fuel. S
   (c) State the possible consequences of operating an engine with a cylinder liner worn beyond normally acceptable limits.

2. *(a)* Describe briefly three methods of crankshaft construction indicating for which type of engine the method is most suitable.
   (b) State the nature of and reasons for the type of finish used at mating surfaces of a shrink fit. S
   (c) Explain:
       (i) Why slippage of a shrink fit can occur.
       (ii) How such slippage may be detected.
       (iii) How slippage may be rectified.

3. During recent months it has been necessary to frequently re-tighten some main engine holding down bolts as the steel chocks have become loose:
   (a) Explain possible reasons for this.
   (b) State with reasons why re-chocking using a different material might reduce the incidence. S
   (c) Explain the possible consequences if the situation is allowed to continue unchecked.

4. *(a)* State, with reasons, why engine air inlet and exhaust passageways should be as large as possible.
   (b) Explain how such passageways can become restricted even when initially correctly dimensioned. S
   (c) Explain the consequences of operating an engine with:
       (i) Restricted air inlet passageways.
       (ii) Restricted exhaust passageways.

5. *(a)* Describe, using sketches if necessary, a main engine chocking system using resin based compounds, explaining how such a system is installed. S

(b) State the advantages and disadvantages of resin based materials for use as chocks when compared with iron or steel.

CHAPTER 3 – CLASS ONE

1. Bunkers have been taken in a port and a sample is sent to a laboratory for analysis. The vessel proceeds to sea before results of the analysis are obtained. The analysis indicates that the fuel is off specification in a number of respects but the fuel must be used as there is insufficient old oil supply available to enable the ship to reach the nearest port. Explain with reasons what action should be taken to minimise damage and enable safe operation of the engine if the following fuel properties were above or below specified levels: S
   (a) Viscosity.
   (b) Compatibility.
   (c) Sulphur.
   (d) Ignition quality.
   (e) Conradson carbon.
   (f) Vanadium and sodium.

2. (a) Describe using sketches a Variable Injection Timing fuel pump and explain how timing is varied whilst the engine is in operation. S
   (b) Explain why it is necessary to adjust the timing of fuel pumps individually and collectively.

3. With respect to residual fuel explain the effects of EACH of the following on engine components, performance and future maintenance, stating any steps which should be taken in order to minimise these effects:
   (a) High Conradson carbon level. S
   (b) Aluminium level of 120 ppm.
   (c) Low ignition quality.
   (d) 450 ppm vanadium plus 150 ppm sodium.

4. (a) With reference to 'slow steaming nozzles' as applied to main engine fuel injectors. State with reasons when and why they would be used. S

(b) State with reasons the engine adjustments required when changing to a fuel having a different ignition quality. Explain the consequences of not making such adjustments.
   (c) State the procedures which should be adopted to ensure that main engine fuel injectors are maintained in good operative order indicating what routine checks should be made.

5. With respect to fuel oil:
   (a) Explain the meaning of term 'ignition quality' and indicate the possible problems of burning fuels of different ignition quality. S
   (b) State how an engine may be adjusted to deal with different fuels of different ignition qualities.
   (c) State how fuel structure dictates ignition quality.

CHAPTER 4 – CLASS ONE

1. *(a)* Sudden bearing failure occurs with a turbocharger which has been operating normally until that point. Explain the possible causes if the turbocharger has:
   (i) Ball or roller bearings. S
   (ii) Sleeve bearings.
   (b) State with reasons the measures to be adopted to ensure that future failure is minimised.

2. With respect to turbochargers indicate the nature of deposits likely to be found on EACH of the following and in each case state the possible consequences of operating with high levels of such deposits and explain how any associated problems might be minimised:
   (a) Air inlet filters.
   (b) Impeller and volute. S
   (c) Air cooler.
   (d) Turbine and nozzles.
   (e) Cooling water spaces.

3. *(a)* State what is meant by the term *surge* when applied to turbochargers. S
   (b) State why surging occurs and how it is detected.

(c) Explain how the possibility of surging may be minimised.

(d) State what action should be taken in the event of a turbocharger surging and explain why that action should not be delayed.

4. It is discovered that delivery of air from a turbocharger has fallen even though engine fuel control has not been changed. State the reasons:
   (a) The causes of such reduced delivery.   S
   (b) The effects of this reduced air supply on the engine.
   (c) The immediate action to be taken.
   (d) How future incidents might be minimised.

5. At certain speed vibration occurs in a turbocharger.
   (a) State with reasons the possible causes.   S
   (b) Explain how the cause can be detected and corrected.
   (c) Explain how the risk of future incidents can be minimised.

CHAPTER 5 – CLASS ONE

1. (a) State the possible reasons for an engine failing to turn over on air despite the fact that there is a full charge of air in the starting air receiver and explain how the problem would be traced.
   (b) Explain how the engine could be started and reversed   S manually in the event of failure of the control system.
   (c) Outline planned maintenance instructions which could be issued to minimise the risk of failure indicated in (a) and (b).

2. Describe the safety interlocks in the air start and reversing system of a main engine.
   What maintenance do these devices require?
   At what interval would they be tested?

3. As Chief Engineer Officer what standing orders would you issue to your engineering staff regarding preparing the main engines for manoeuvring?

4. Routine watchkeeping reveals that a cylinder air start valve is leaking.
   What are the dangers of continued operation of the engine?
   What steps would you take if the vessel was about to commence manoeuvring?

5. Describe the main engine shutdown devices. How and how often would you test them?
   The shut down system on the main engine fails, immobilising the engine. Checks reveal that all engine operating parameters are normal. What procedures would you, as Chief Engineer Officer, adopt to operate the unprotected engine to enable the vessel to reach port?

CHAPTER 6 – CLASS ONE

1. *(a)* Describe briefly the operation of an electrical or hydraulic main engine governor.
   (b) For the type described indicate how failure can occur and the action to be taken if immediate correction cannot be achieved and the engine must be operated.   S

2. Complete failure of the UMS, bridge control and data logging systems has occurred resulting in the need for the main engine to be put on manual control and monitoring:
   (a) State with reasons six main items of data which require to be monitored and recorded manually.
   (b) Explain how a watchkeeping system should be arranged to provide for effective monitoring and control of the main engine.   S
   (c) Explain how the staff will be organised to allow the engine to be manoeuvred safely and state the items of plant which will require attention during such manoeuvring.

3. Discuss the relative merits and demerits of hydraulic and electronic main engine governors.

4. Describe, with the aid of a block diagram, a bridge control system for main engine operation.

As Chief Engineer Officer what standing orders would you issue to your engineering staff when the vessel was operating under bridge control?

5. Describe a jacket cooling water system temperature controller. When operating under low load conditions for an extended period how can cylinder liner corrosion be minimised?

CHAPTER 7 – CLASS ONE

1. During a period of manoeuvring it is noticed that difficulty is being experienced in maintaining air receiver pressure:
   (a) State, with reasons, possible explanations.
   (b) Explain how the cause may be traced and rectified.   S
   (c) State what immediate action should be taken to ensure that the engine movements required by the bridge are maintained.

2. (a) Explain why it is essential to ensure adequate cooling of air compressor cylinders, intercoolers and aftercoolers.
   (b) State, with reasons, the possible consequences of prolonged   S operation of the compressor if these areas are not adequately cooled.

3. (a) With reference to air receivers explain:
       (i) Why regular internal and external inspection is advisable.
       (ii) Which internal areas of large receivers should receive particularly close attention.   S
       (iii) How the internal condition of small receivers is checked.
   (b) Where significant corrosion is found during an internal inspection what factors would you take into account when revising the safe working pressure?

4. It has been found that during recent periods of manoeuvring a number of air start valve bursting discs or cones have failed:
   (a) Explain the possible reasons for this.
   (b) Indicate how the actual cause might be: S
       (i) Detected.
       (ii) Rectified.

5. (a) State why starting air compressor performance deteriorates in service and how such deterioration is detected. S
   (b) Explain the dangers associated with some compressor faults.

CHAPTER 8 – CLASS ONE

1. (a) Explain the advantages and problems of using aluminium in the construction of composite pistons for medium speed engines.
   (b) Briefly describe the removal, overhaul and replacement of a pair of pistons connected to a single crank of a vee-type engine, explaining any problems regarding the bottom end bearings.

2. (a) Explain the advantages of fitting highly rated medium speed engines with double exhaust and air inlet valves.
   (b) State the disadvantages of double valve arrangements. S
   (c) Explain the possible causes of persistent burning of exhaust valves if it is:
      (i) General to most cylinders.
      (ii) Specific to a single cylinder.

3. Explain the problems associated with medium speed diesel exhaust valves when operating with heavy fuel oil.
   How can these problems be minimised?
      (i) By design.
      (ii) By maintenance.

4. Describe a suitable maintenance schedule for one unit of a medium speed diesel engine operating on heavy oil?

5. Describe the torsional vibration of medium speed diesel engine crankshafts.
   Describe, with the aid of sketches, a coupling that will aid the damping of torsional vibration.

CHAPTER 9 – CLASS ONE

1. As Chief Engineer Officer, what standing orders would you issue your engineering staff to ensure that the auxiliary boiler was operated in a safe and efficient manner?

2. Describe a waste heat plant that is able to produce sufficient steam to a turbo-generator to supply the entire ship's electrical load at sea.
   Due to trading requirements the vessel is sailing at reduced speed.
   Describe the steps you would take to ensure the slowest ship's speed commensurate with supplying sufficient steam to the turbo-generator without allowing the boiler to fire or starting diesel generators.

3. Describe, with the aid of sketches, an auxiliary boiler suitable for use with a waste heat unit.
   Explain how the pressure of the steam plant is maintained when operating under low steam load conditions.

4. Sketch and describe a composite thimble tube boiler. Describe how the thimble tubes are fitted and discuss burning of tube ends and other possible defects.

5. You are Chief Engineer Officer of a motor vessel equiped with a steam plant incorporating a waste heat unit in the engine uptake. On passage it is reported to you that the uptake temperature is rising.
   (a) What would this information indicate and what steps would you take?
   (b) How could you prevent a reoccurrence?

CHAPTER 1 – CLASS TWO

1. *(a)* State the ideal cycle most appropriate to the actual operations undergone in the modern diesel engine.
   (b) Give reasons why the actual cycle is made approximate to the ideal heat exchange process.
   (c) State how the combustion process in the actual cycle is made approximate to the ideal heat exchange process.

2. *(a)* State why bottom end bolts of 4-stroke engines are susceptible to failure.
   (b) Sketch a bottom end bolt of suitable design.
   (c) Explain how good design reduces possibility of failure. S
   (d) State how the possibility of failure is reduced by good maintenance.
3. *(a)* Explain why in large, slow speed engines, power balance between cylinders is desirable.
   (b) State why it is never achieved in practice.
   (c) Describe how power balance between cylinders of a medium speed engine is improved. S
   (d) Describe how power balance in a slow speed engine is improved.

4. *(a)* Give an example of each of the four types of 2-stroke engine indicator diagrams, explain how each is taken and the use to which it is put.
   (b) Illustrate two defects which can show up on a compression card. S
   (c) How is cylinder power balance checked on a higher speed engine?

5. *(a)* Explain how the power developed in an engine cylinder is determined:
       (i) From indicator cards. S
       (ii) By electronic means.
   (b) State which of these is the most representative and why.

CHAPTER 2 – CLASS TWO

1. *(a)* State TWO reasons why large crankshafts are of semi-built construction.
   (b) State SIX important details of crankshaft construction that will reduce the possibility of fatigue failure. S
   (c) List FOUR operational faults that may induce failure in a crankshaft.

2. *(a)* State the nature of the stresses to which crank webs of large diesel engines are subjected. S
   (b) Explain how they are designed and manufactured to resist these stresses.

3. (a) State the reason for fitting crosshead guides to engines and explain why 'ahead' and 'astern' faces are required with uni-directional engines.   S
 (b) Describe how crosshead guide clearance is checked and adjusted.
 (c) List reasons for limiting such crosshead clearance.

4. (a) State why bedplates of large engines are fitted with chocks rather than directly on foundation plates.
 (b) Sketch an arrangement of lateral chocking showing the position relative to the engine.   S
 (c) State why such an arrangement is employed.
 (d) State the factors that determine the spacing of the main chocks.

5. With reference to auxiliary diesel engine machinery:
 (a) (i) State why this may be mounted on resilient mountings.
 (ii) State why such mountings have great flexibility.   S
 (b) State why limit stops are provided.
 (c) State how the external piping is connected.

CHAPTER 3 – CLASS TWO

1. (a) Describe, with the aid of sketches, a fuel pump capable of variable injection timing.
 (b) State why injection timing might need to be changed.   S
 (c) State how injection timing is adjusted whilst the engine is running.

2. (a) Sketch and describe a fuel valve for a diesel engine.
 (b) State FOUR factors which indicate that fuel valve(s) require attention.   S

3. (a) State the factors that influence:
 (i) Droplet size during fuel injection.
 (ii) Penetration.   S
 (b) State TWO methods of improving air turbulence.

4. *(a)* Sketch and describe a jerk type fuel pump that is not helix controlled.
   (b) Explain how the pump may be timed.   S
   (c) State TWO advantages of this type of pump.

5. *(a)* Sketch a main engine fuel pump of the scroll type.
   (b) Explain how the fuel quantity and timing are adjusted.   S
   (c) To what defects is this type of pump subject and how is the pump adjusted to counter their effects?

CHAPTER 4 – CLASS TWO

1. *(a)* Describe with the aid of sketches:
      (i) A pulse turbocharger system.
      (ii) A constant pressure turbocharger system.
   (b) State the advantages and disadvantages of each system in   S
   Q.1.*(a)*. for use with marine propulsion engines.
   (c) In the event of turbocharger failure with one of the systems in Q.1.*(a)*. state how the engine could be arranged to operate safely.

2. *(a)* Sketch a simple valve timing diagram for a naturally aspirated 4-stroke engine.
   (b) Sketch a simple valve timing diagram for a supercharged 4-   S
   stroke engine.
   (c) Comment on the differences between the two above diagrams.

3. *(a)* Sketch and describe a turbocharger with a radial flow gas turbine showing the position of the bearings.   S
   (b) State the advantages of radial flow gas turbines.

4. *(a)* State why turbochargers are used to supply air to an engine rather than expanding the gas further in the cylinder and then employing crank driven scavenge pumps.
   (b) Explain what measures should be adopted to ensure safe   S
   operation of the engine should all turbochargers be put out of action.

(c) State why a 2-stroke cycle engine relies upon a pressurised combustion air supply but a 4-stroke cycle engine does not.

5. (a) Explain why air coolers and water separators are fitted to large turbocharged engines. S
 (b) Sketch a water separator, explain how it operates and indicate its positioning in the engine.
 (c) What are the defects to which coolers and separators are susceptible?

CHAPTER 5 – CLASS TWO

1. (a) Sketch a starting air distributor used for a large reversible engine.
 (b) Explain how the engine may be started with the crankshaft in any rotational position. S
 (c) Explain how the engine is started on air in either direction.

2. (a) Sketch a pneumatically operated starting air valve.
 (b) Explain how the valve is operated. S
 (c) State what normal maintenance is essential and the possible consequence if it is neglected.

3. (a) Sketch and describe the reversing system for a large slow speed diesel engine. S
 (b) List the safety devices fitted to the air start system.

4. (a) Explain why it is necessary to have air start overlap.
 (b) Show how air start timing is affected by exhaust timing. S
 (c) State why the number of cylinders have to be taken into consideration.

5. (a) Sketch an engine air start system from the air receiver to the cylinder valves and describe how it operates. S
 (b) List the safety devices and interlocks incorporated in such a system and state the purpose of each.

CHAPTER 6 – CLASS TWO

1. With reference to a jacket water temperature control system:
   (a) Sketch and describe such a system:
   (b) (i) Explain how disturbances in the system may arise.   S
       (ii) Describe how these disturbances may be catered for.

2. *(a)* Construct a block diagram, in flow chart form, to show the sequence of operations necessary for the starting of a diesel engine on bridge control.   S
   (b) Identify the safety features incorporated in the system of Q.2.*(a)*.

3. *(a)* Sketch a cylinder relief valve suitable for a large engine.
   (b) State with reasons why such a device is required.
   (c) If the relief valve lifts state the possible causes and indicate the rectifying action needed to prevent engine damage.   S
   (d) State why the relief valve should be periodically overhauled even though it may never have lifted.

4. With reference to mechanical/hydraulic governors explain:
   (a) Why the flyweights are driven at a higher rotational speed than the engine.
   (b) How dead band effects are reduced.   S
   (c) How hunting is reduced.
   (d) How the output torque is increased.

5. Sketch and describe a hydraulic governor with proportional and reset action.

CHAPTER 7 – CLASS TWO

1. *(a)* Sketch a jacket water cooling system.
   (b) State why chemical treatment of the jacket cooling water is necessary.   S
   (c) Describe how the correct concentration of the chemicals in the jacket water cooling system may be determined.

2. *(a)* Explain how oil may become mixed with starting air and state the attendant dangers.
   (b) Describe how this contamination may be reduced or prevented.   S

(c) Outline the dangers of lubricating oil settling in air starting lines.

(d) How may an air start explosion be initiated?

3. *(a)* Explain why air compression for starting air duties is carried out in stages and why those stages are apparently unequal. S

    (b) What is the purpose of an intercooler and explain why it is important that it is kept in a clean condition?

    (c) What is the significance of clearance volume to compressor efficiency?

    (d) What is bumping clearance and how is it measured?

4. *(a)* State why compressor suction and delivery valves should seat promptly.

    (b) Explain the effect on the compressor if the air is induced into the cylinder at a temperature higher than normal. S

    (c) What would be the effect of the suction valves having too much lift.

    (d) Explain why pressure relief devices are fitted to the water side of cooler casings.

5. *(a)* State why inhibitors are employed with engine cooling water even though distilled water is used for that purpose.

    (b) State the merits and demerits of the following inhibitors used in engine cooling water systems: S
      (i) Chromate.
      (ii) Nitrite-borate.
      (iii) Soluble oil.

    (c) Briefly explain how each inhibitor functions.

CHAPTER 8 – CLASS TWO

1. Describe, with the aid of sketches, an exhaust valve of a medium speed diesel engine suitable for use with heavy fuel oil. Explain the procedure adopted when overhauling this valve.

2. Describe, with the aid of sketches, a piston suitable for use in a medium speed engine. Why is aluminium being generally superceded for pistons on highly rated medium speed engines.

3. Describe with the aid of sketches a system for main propulsion in which two medium speed diesel engines are coupled to a single propeller.

4. Describe the advantages and disadvantages of medium speed diesel engines compared to large slow running engines.

5. Explain why lubricating oil consumption is greater in medium speed engines than in slow running diesels and the steps taken to minimise the consumption.

CHAPTER 9 – CLASS TWO

1. *(a)* Describe, with the aid of sketches, an arrangement for producing electricity using steam generated from waste heat.
   (b) State how electricity can be generated with the system in Q.1.*(a)* when the engine is not operating.
   (c) State the circumstances which could lead to an emergency shut-down of the steam plant in Q.9.*(a)* and the use of diesel engines for electrical generation.

2. Describe the inspection of an auxiliary boiler.
   What precautions should be taken prior to entering the boiler?

3. Describe, with the aid of sketches a boiler which may be alternatively fired or heated with main engine exhaust gas in which the heating surfaces are common. Describe the change over arrangements and state any safety devices fitted to this gear.

4. What are the precautions that should be taken before and during the "flashing up" operation of an auxiliary boiler?
   State the checks carried out on the boiler when a fire is established.

5. What are the advantages and disadvantages of forced circulation and natural circulation multi-boiler installations?
   How can the steam pressure of the waste heat plant be controlled when operating on exhaust gas?

6. Describe the dangers of dirty uptake in the waste heat unit. Explain how these dangers are minimised.

SPECIMEN QUESTIONS – CLASS ONE

1. *(a)* Define the term *hot spot*.
   (b) State SIX specific areas in a diesel engine where hot spots have occurred. S
   (c) State other factors that may contribute to the occurrence of a crankcase explosion.

2. With reference to crankcase explosions state:
   (a) The conditions that may initiate an explosion.
   (b) What may cause a secondary explosion. S
   (c) How a crankcase explosion relief valve works.

3. *(a)* State the basic processes leading up to a crankcase explosion and explain how a secondary explosion can occur. S
   (b) List with reasons the precautions which can be taken to minimise the risk of a crankcase explosion occurring.

4. *(a)* Explain how a primary crankcase explosion is caused and how it may trigger a secondary explosion.
   (b) Indicate the possible benefits or dangers of the following features on the likely development of a crankcase explosion: S
   (i) Oil mist detector.
   (ii) Inert gas injection.
   (iii) Infra-red heat detectors.
   (iv) Bearing shells having a layer of bronze between the white metal and steel backing steel.

5. *(a)* Describe, using sketches if necessary, the procedure for complete inspection of a propulsion engine main bearing and journal.
   (b) State the possible bearing and pin defects which might be encountered. S
   (c) State what precautions should be taken before returning an engine to service following such bearing inspection and adjustment.

6. *(a)* Explain the reason for fitting crossheads and guides to large slow speed engines.
   (b) Explain:
      (i) Why guide clearance is limited.   S
      (ii) How guide clearance is adjusted.
      (iii) How guide alignment is checked.

7. *(a)* During an inspection it is noticed that tie rods of certain main engine units have become slack, state with reasons the possible causes of this.
   (b) Explain how correct tension is restored and the risk of future slackness minimised.   S
   (c) A tie rod has fractured and cannot be replaced immediately. State with reasons the course of action to be adopted in order to allow the engine to be operated without further damage.

8. *(a)* Explain the term *fuel ignition quality* and indicate how a fuel's chemical structure influences its value.
   (b) State, with reasons, the possible consequences of operating an engine on a fuel with a lower ignition quality than that for which it is timed.   S
   (c) (i) Explain how an engine might be adjusted to burn fuel of different ignition quality.
      (ii) State what checks can be carried out in order to determine that the engine is operating correctly.

9. *(a)* Describe the phenomenon of surging as applied to turbochargers.
   (b) Explain why turbochargers are not designed to completely eliminate the possibility of surging.   S
   (c) State with reasons the possible consequences of allowing a turbocharger to continue to operate whilst it is surging.

10. With reference to turbocharger systems state how deposit build-up might be detected on the following parts and explain the consequences on turbocharger and engine operation of excessive deposits: S
    (a) Suction filter.
    (b) Impeller.

(c) Turbine nozzle and blades.
(d) Air cooler.

11. Difficulty is experienced in starting an engine even though there is full air pressure in the air receivers and fuel temperature is correct. Explain how the cause of the problem can be: S
    (a) Detected.
    (b) Rectified.

12. With reference to piston ring and liner wear:
    (a) State, with reasons, the causes of abnormal forms of wear known as cloverleafing and scuffing (microseizure);
    (b) Explain how cylinder lubrication in terms of quantity and quality can influence wear;   S
    (c) Describe the procedure for determining whether piston rings are suitable for use.

13. With reference to main engine holding down studs/bolts:
    (a) Explain the causes of persistent slackening.
    (b) State, with reasons, the likely consequences of such slackening.   S
    (c) Describe how future incidents of slackening might be minimised.

14. (a) Inspection of an engine indicates an unexpected increase in cylinder liner wear rate, state with reasons the possible causes if:
    (i) The problem is confined to a single cylinder.
    (ii) The problem is common to all cylinders.   S
    (b) Explain how cylinder wear rate may be kept within desired limits and indicate the instructions to be issued to ensure that engine room staff are aware as to how this can be achieved.

15. Cracks have been discovered between the crankpin and web on a main engine crankshaft:
    (a) Describe action to be taken in order to determine the extent of the cracking.   S

(b) Explain the most likely reasons for the cracking.
(c) State, with reasons, the action to be taken in order that the ship may proceed to a port where thorough inspection facilities are available.

16. It is found that tie rods are persistently becoming slack:
    (a) State, with reasons, the possible causes.
    (b) State, with reasons, the likely effects on the engine if it is allowed to operate with slack tie rods. S
    (c) Explain how this problem can be minimised.

17. As Chief Engineer Officer, explain the procedure to be adopted for the complete inspection of a main engine cylinder unit emphasising the areas of significant interest. S

18. *(a)* The water jacket on a turbocharger casing has fractured allowing water into the turbine side. State possible reasons for this.
    (b) Explain how the engine may be kept operational and the restrictions now imposed upon the operating speed. S
    (c) State how the fracture can be rectified and how future incidents can be minimised.

19. *(a)* State the conditions which could result in a fire in the tube space and/or uptakes of a waste heat boiler.
    (b) State how such conditions can occur and how the risk of fire can be minimised. S
    (c) State how such fires can be dealt with.

20. As Chief Engineer, explain the procedure to be adopted for the survey of an air compressor on behalf of a classification society. S

21. *(a)* Identify, with reasons, the causes and effects of misalignment in large, slow speed, engine crankshafts.
    (b) Describe how the alignment is checked. S
    (c) State how the measurements are recorded and checked for accuracy.

22. *(a)* Explain why side and end chocking arrangements are provided for large direct drive engines.
    (b) State, with reasons, why non-metallic chocking is considered superior to metallic chocking. S
    (c) State why top bracing is sometimes provided for large engines and explain how it is maintained in a functional condition.

23. *(a)* As Chief Engineer, describe how a complete inspection of a main engine turbocharger may be carried out indicating, with reasons, the areas requiring close attention. S
    (b) Describe defects which may be found during inspection and their possible cause.

24. The main engine has recently suffered problems related to poor combustion and inspection indicates that a number of injector nozzles are badly worn:
    (a) Explain the possible causes of the problem and how they may be detected. S
    (b) State how future problems of a similar nature can be minimised.

25. With reference to fuel pumps operating on residual fuel:
    (a) (i) State, with reasons, the defects to which they are prone.
        (ii) Explain the effects of such defects on engine performance. S
    (b) State, with reasons corrective action necessary to restore a defective fuel pump to normal operation.
    (c) Suggest ways in which the incidence of these defects might be minimised.

SPECIMEN QUESTIONS – CLASS TWO

1. Describe the routine maintenance necessary on the following components in order to obtain optimum performance from a main engine turbocharger: S
   (a) Lubricating oil for ball bearings.
   (b) Air intake silencer/filter.

(c) Turbine blades.
(d) Diffuser ring.

2. (a) List the advantages of multi-stage air compression with intercooling compared with single stage compression.
   (b) Explain the faults which may be encountered during overhaul of the H.P. stage and indicate how they may be rectified.   S

3. (a) Outline the problems associated with air compressor cylinder lubrication indicating why it should be kept to a minimum.
   (b) State why a restricted suction air filter might make the situation worse and lead to the possibility of detonation in the discharge line.   S
   (c) Explain why the compressor discharge line to the air receiver should be as smooth as possible with the minimum number of joints and connections.

4. (a) Explain the need for additives in engine jacket water cooling systems.
   (b) State what factors determine the choice of chemicals used.   S
   (c) State why chromates are seldom used.

5. (a) Give a simple line sketch of a jacket water cooling system.
   (b) Describe a control system capable of maintaining the jacket water temperature within close limits during wide changes in engine load.   S

6. (a) Sketch an arrangement for securing turbocharger blades to the blade disc.
   (b) How is blade vibration countered?
   (c) What is the cause of excessive turbocharger rotor vibration?   S
   (d) Briefly describe an in-service cleaning routine for the gas side of a turbocharger.

7. (a) Describe with sketches a scroll type fuel pump.
   (b) Explain how the quantity of fuel is metered and how the governor cut out functions.   S

(c) State how this type of pump is set after overhaul.
(d) State the reasons that necessitate pump overhaul.

8. *(a)* Sketch a fuel injector.
   (b) Explain how it operates and what determines the point at which injection occurs.   S
   (c) Describe the defects to which injectors are prone.
   (d) How can injection be improved when a low speed engine is to operate at prolonged low load?

9. With reference to turbocharging:
   (a) (i) Explain the terms pulse system and constant pressure system.
   (ii) List the advantages of each.
   (b) State how in a pulse system the exhaust from one cylinder may be prevented from interfering with the scavenging of another.   S
   (c) State why electrically driven blowers are usually fitted in addition to turbochargers.

10. *(a)* Show how combustion forces are transmitted to the cross members of the bedplate.   S
    (b) Describe TWO means by which the stresses within the cross members can be accommodated.

11. *(a)* Describe how crankshaft alignment is checked.
    (b) Identify, with reasons the causes of crankshaft misalignment.   S
    (c) State how the measurements are recorded.

12. *(a)* Sketch a cross-section of a main engine structure comprising bedplate, frames and entablature showing the tie bolts in position.
    (b) Explain why tie bolts need to be used in some large, slow speed engines.   S
    (c) Explain in detail how the tie bolts are tensioned.

13. Give reasons why, when compared to the other bearings of large slow speed engines, top end bearings: S

(a) Are more prone to failure.
(b) Have a greater diameter in proportion to pin length.

14. (a) State how engine cylinder power is checked and approximate power balance is achieved.
    (b) Explain why the methods of checking may differ between slow and higher speed engines.   S
    (c) State why perfect cylinder power balance cannot be achieved.
    (d) State the possible engine problems resulting from poor cylinder power balance.

15. (a) Describe with sketches the mono-box frame construction which is being used to replace the traditional A-frame arrangement for some crosshead engines.   S
    (b) State why this form of construction is considered to be more suitable than one using A-frames.

16. (a) State TWO reasons why large crankshafts are of semi-built or fully built construction.
    (b) State SIX important details of crankshaft construction that will reduce the possibility of fatigue failure.   S
    (c) State FOUR operational faults that may induce fatigue failure.

17. (a) Define the cause of corrosive wear on cylinder liners and piston rings.
    (b) Explain the part played by cylinder lubrication in neutralising this action.   S
    (c) State how the timing, quantity and distribution of cylinder oil is shown to be correct.

18. With reference to large fabricated bedplates give reasons to explain:
    (a) Why defects are likely to occur in service and where they occur.   S
    (b) How these defects have been avoided in subsequent designs.

19. (a) Define the cause of cylinder liner and piston ring wear.   S

(b) Describe how cylinder liner wear is measured and recorded.

(c) Explain the possible consequences of operating a main engine with excessive cylinder liner wear.

20. *(a)* Sketch a main engine holding down arrangement employing long studs and distance pieces.

(b) Explain why the arrangement sketched in Q.6.*(b)*. may be employed in preference to short studs.   S

(c) Describe, with the aid of sketches, how transverse movement of the bedplate is avoided.

21. *(a)* Briefly discuss the relative advantages and disadvantages of oil and water for cooling.

(b) Sketch a piston for a large two stroke crosshead engine indicating the coolant flow.   S

(c) State the causes of piston cracking and burning, and how it can be avoided.

22. *(a)* Sketch the arrangement of a large 2-stroke engine cylinder liner in position in the cylinder block.

(b) Describe how jacket water sealing is accomplished between liner and cylinder block.   S

(c) For the liner chosen illustrate the directions of cooling water flow, exhaust gas flow and combustion air flow.

(d) Explain how thermal expansion of the liner is accommodated.

23. *(a)* Give the reasons for progressive 'fall-off' of piston ring performance in service.

(b) State, with reasons, which ring clearances are critical.   S

(c) State what effects face contouring, bevelling, ring cross section and material properties of rings and liners have on ring life.

24. *(a)* Sketch the arrangement for connecting a piston to the crosshead.

(b) State the type of piston coolant employed and show how the coolant is directed to and from the piston.   S

(c) State the precautions to be exercised when lifting or overhauling the piston described.

25. *(a)* Explain the reasons for employing two air inlet and two exhaust valves for high powered trunk piston 4-stroke engines.

(b) State the problems relating to tappet setting with such valves.

(c) Sketch a caged valve as fitted to a trunk piston engine.

S

INDEX

A
"A" Frames 38
Abrasive Wear 76
Adiabatic Compression 195
" Operation 1
Air Coolers 139
Air Standard Efficiency 8
Air Vessels 208
Aluminium Contamination 127
Atomisation 101

B
Bearing Corrosion 58
Bedplate 30
Bending Moment 35
Bore Cooling 70,96
Bosch Fuel Pump 113
" Operation 112
Brake Thermal Efficiency 2

C
Carnot Cycle 9
Catalytic Fines 127
Cloverleafing 78,80
Cochran Composite Boiler 247
Common Rail Injection 106
Compression Diagram 14
" Ratio 3
Compressor Valve Defects 203
Compressor Valves 204
Constant Pressure Turbocharger ...137
Cooling Systems 209
Corrosive Wear 76
Crankcase Explosion 257
" Protection Devices 259
" Safety Arrangements ...259
Crankshaft 46
" Deflections 60
" Fillet Radius 53
" Journal Reference Marks ...53
" Misalignment 59
" Shrink Fitting Slippage 54
" Stresses 49
" Torsional Vibration 57
Crankshaft: one piece 50
Cylinder Liner 70
" Wear 75

Cylinder Liner Wear Profile 79
Cylinder Lubrication 72

D
Dew Point 245
Diesel Cycle 9
Diesel Knock 111
Dissociation 11
Draw Card 13
Dual Cycle 9
Dual Pressure Multiboiler System...253

E
Electronic Indicating Equipment 24
Embedded Crankshaft 30
Epoxy Resin Engine Chocks 43
Exhaust Blowdown 132
Exhaust Gas Analysis 242
Exhaust Gas Power Turbine 262
Exhaust Valve 224
" Rotocap 226
Explosion Door 259

F
Fatigue 26
" Limit 26
Flame Trap 260
Flexible Coupling 222
Fluid Coupling 219
Forced Circulation Waste Heat Plant ...251
Forged Crankshaft 50
Four Stroke Engine Cylinder Head ...97
Four Stroke Engine Valve Actuation ...99
Four Stroke Piston 85
Four-Stroke Engine Cylinder Head ...96
Fresh Water Cooling 213
Fretting Corrosion 58
Friction Clutch 221
Fuel Air Ratio 3
Fuel Analysis 125
Fuel Cooler 124
Fuel Management 125

G
Gas Turbine 262
" Cycle 10
Geislinger Coupling 223

H

Heat Balance 4,5
Heat Effected Zone 33
Holding Down Arrangements 38
Hydraulic Exhaust Valve........................96
Hydraulic Fuel Injector 105
Hydrodynamic Librication 67

I

Ideal Cycles 6
Impingement 101
Incompatibility 127
Indicator Diagram 14
Industrial Gas Turbine 264
Instability 127
Isochronous 183
Isothermal Operation...............................1

J

Jerk Injection......................................106
Joule Cycle 9

L

Light Spring Diagram........................14,16
Load Controlled Cylinder Cooling 77,210
Load Diagram 6
Long Sleeve Holding Down Bolt 41
Loop Scavenge 2
Lubricating Oil Additives 72
Lubricating Oil Analysis 65
Lubricating Oil Cooling 213

M

Medium Speed Diesels 217
Membrane Wall...................................239
Micro-seizure 78
Monoblock Frames 38

N

Natural Circulation Waste Heat Plant 250

O

Oil Mist Detector 260
Oil cooled piston 83
Otto Cycle 6

P

Package Boilers..................................237
Penetration...101

Performance Curves 3
Pilot Injection 123
Piston Cooling Control 189
Piston Ring Defects 90
Piston Ring Manufacture.......................90
Piston Ring Profile 88
Piston Rings...87
Pistons 78
Polytropic Operation 1
Pour Point 124
Power Card 13
Pressure Charging 133
Proportional Action 182

R

Radial Flow Turbine 146
Resilient Engine Mountings 45
Reset Action 182
Reverse Reduction Gear 220
Rigid Engine Foundations 39

S

Secondary Explosion...........................258
Selective Catalytic Reduction (SCR) 267
Sulphuric Acid 245
Specific Fuel Consumption 3
Speed Droop 179
Starting Air Distributor 163
Starting Air Overlap 159
Starting Air Valves 161
Starting Interlocks 172
Stellite 224
Sulzer Valve Type Fuel Pump 118
Sunrod Boiler 237
Sunrod Element 239
Superheater 253
Surging 151

T

Theoretical Engine and Cycles 9
Thimble Tube Boiler 248
Three Stage Air Compressor 200
Tie-Rod Tensioning...............................36
Timed Injection 106
Timing Diagram 17
Turbocharger Dry Cleaning 151
Turbocharger Waterwash 150
Turbochargers 141
Turbulence...101

Two Stage Air Compressor 202
Two Stage Fuel Injection 120

U

Unattended Machinery Space 194
Uncooled Turbo Charger 144
Underslung Crankshaft 30
Uniflow Scavenge 2
Unloading Valve 207

V

Vanadium 127,224
Vapour Vertical Boiler 240
Variable Injection Timing 109
Viscosity 102
Volumetric Efficiency 1,197

W

Waste Heat Boilers 245
Waste Heat Recovery 248
Water Separator..................................140
Watercooled piston 82
Welded Crankshaft 54
Welded Crankshaft 55

REED'S MARINE ENGINEERING SERIES

Vol. 1   MATHEMATICS
Vol. 2   APPLIED MECHANICS
Vol. 3   APPLIED HEAT
Vol. 4   NAVAL ARCHITECTURE
Vol. 5   SHIP CONSTRUCTION
Vol. 6   BASIC ELECTROTECHNOLOGY
Vol. 7   ADVANCED ELECTROTECHNOLOGY
Vol. 8   GENERAL ENGINEERING KNOWLEDGE
Vol. 9   STEAM ENGINEERING KNOWLEDGE
Vol. 10 INSTRUMENTATION AND CONTROL SYSTEMS
Vol. 11 ENGINEERING DRAWING
Vol. 12 MOTOR ENGINEERING KNOWLEDGE

REED'S ENGINEERING KNOWLEDGE FOR DECK OFFICERS
REED'S MATHS TABLES AND ENGINEERING FORMULAE
REED'S MARINE DISTANCE TABLES
REED'S OCEAN NAVIGATOR
REED'S SEXTANT SIMPLIFIED
REED'S SKIPPERS HANDBOOK
REED'S COMMERCIAL SALVAGE PRACTICE
REED'S MARITIME METEOROLOGY
SEA TRANSPORT – OPERATION AND ECONOMICS

These books are obtainable from all good Nautical Booksellers
or direct from:

THOMAS REED PUBLICATIONS
The Barn, Ford Farm
Bradford Leigh
Bradford-on-Avon
Wiltshire BA15 2RP
United Kingdom

Tel: 01225 868821
Fax: 01225 868831

Email: tugsrus@abreed.demon.co.uk